Springer Theses

Recognizing Outstanding Ph.D. Research

Aims and Scope

The series "Springer Theses" brings together a selection of the very best Ph.D. theses from around the world and across the physical sciences. Nominated and endorsed by two recognized specialists, each published volume has been selected for its scientific excellence and the high impact of its contents for the pertinent field of research. For greater accessibility to non-specialists, the published versions include an extended introduction, as well as a foreword by the student's supervisor explaining the special relevance of the work for the field. As a whole, the series will provide a valuable resource both for newcomers to the research fields described, and for other scientists seeking detailed background information on special questions. Finally, it provides an accredited documentation of the valuable contributions made by today's younger generation of scientists.

Theses are accepted into the series by invited nomination only and must fulfill all of the following criteria

- They must be written in good English.
- The topic should fall within the confines of Chemistry, Physics, Earth Sciences, Engineering and related interdisciplinary fields such as Materials, Nanoscience, Chemical Engineering, Complex Systems and Biophysics.
- The work reported in the thesis must represent a significant scientific advance.
- If the thesis includes previously published material, permission to reproduce this must be gained from the respective copyright holder.
- They must have been examined and passed during the 12 months prior to nomination.
- Each thesis should include a foreword by the supervisor outlining the significance of its content.
- The theses should have a clearly defined structure including an introduction accessible to scientists not expert in that particular field.

More information about this series at http://www.springer.com/series/8790

Stefan Putz

Circuit Cavity QED with Macroscopic Solid-State Spin Ensembles

Doctoral Thesis accepted by
TU Wien, Vienna, Austria

 Springer

Author
Dr. Stefan Putz
Department of Physics
Princeton University
Princeton, NJ
USA

Supervisor
Dr. Johannes Majer
TU Wien, Atominstitut
Vienna
Austria

ISSN 2190-5053 ISSN 2190-5061 (electronic)
Springer Theses
ISBN 978-3-319-88247-5 ISBN 978-3-319-66447-7 (eBook)
DOI 10.1007/978-3-319-66447-7

Printed on acid-free paper

This Springer imprint is published by Springer Nature
The registered company is Springer International Publishing AG
The registered company address is: Gewerbestrasse 11, 6330 Cham, Switzerland

To Stephanie

Supervisor's Foreword

Hybrid quantum systems have gained a lot of interest as future quantum technologies. The hope is that two different quantum systems can be combined and that the resulting system profits from advantages of the individual systems. Therewith one intends to pool the strength of different technologies and realize improved quantum devices. The thesis presented by Stefan Putz aims at exploring hybrid quantum devices using superconducting circuits and spin systems, in particular nitrogen-vacancy color centers in diamond. He tried to combine the advantages of superconducting devices, like low losses and exibility with the benefits of the nitrogen-vacancy center, such as the long coherence times and confinement in a host crystal.

In the first two sections, Stefan Putz lays out the theoretical foundations for the thesis. Therein he summarizes the relevant literature and illustrates the relevant physics with analogies like the mass on a spring system or the LC circuit. One of the major results of the thesis is the experimental demonstration of the cavity protection effect. Stefan Putz studied the dynamics of a superconducting cavity strongly coupled to an ensemble of nitrogen-vacancy centers in diamond. He experimentally observed how decoherence induced by inhomogeneous broadening can be suppressed in the strong-coupling regime, a phenomenon that is now known as "cavity protection". Continuing on this project, Stefan Putz developed a scheme to engineer long-lived coherent states in a hybrid system. This scheme was successfully implemented and improved the coherence times by orders of magnitude. Remarkably, this strongly coupled system performs better than the two individual subsystems. Therefore, this experiment is the first one to live up to the promise of hybrid quantum systems and marks a milestone in the field.

In Chap. 7 Stefan Putz demonstrates that hybrid systems can also be used to study nonlinear phenomena. In particular, the hybrid system shows amplitude bistability and allows observing the time dynamics conveniently. He demonstrates a critical slowing down of the cavity population on the order of several tens of thousands of seconds—a timescale much longer than observed so far for this effect. The thesis of Stefan Putz presents several important advances in the field of hybrid quantum systems. Therefore, I am convinced that it will be an important reference for future experiments and novel quantum devices.

Vienna, Austria   Johannes Majer
July 2017

Abstract

Hybrid quantum systems are one of the most promising implementations of future quantum computation, communication, and simulation devices. These technologies will have a major impact on society and industry and herald the start of a new age. However, in the early stages of this development, crucial technological limitations have to be addressed and solved. As one part of this giant puzzle, this thesis examines the basic phenomena occurring when interfacing macroscopic spin ensembles with a single-mode cavity. Understanding these underlying principles is key in the possible implementation of quantum memories based on solid-state spin ensembles.

The hybrid solid-state quantum system studied in this thesis consists of a superconducting microwave cavity strongly coupled to an ensemble of electron spins hosted by nitrogen-vacancy centers in diamond. One of the main results of the experiments carried out demonstrates how the total decoherence rate scales in these systems. As is shown the collective enhanced coupling strength allows the suppression of spin dephasing induced by inhomogeneous spin broadening. This effect is known as "cavity protection effect" and the total decoherence rate scales with the collective cavity spin interaction strength.

The hybridization acts beneficially on the system coherence time. However, these times can be drastically improved by spectral hole burning, beyond the natural limit attainable by the "cavity protection effect". The observed long-lived coherence truly lives up to the promise of hybrid systems to perform better than its individual subcomponents. This demonstrates that dark states can be used to coherently exchange energy between the cavity and spin ensemble. These engineered dark states are used to induce coherent Rabi oscillations and a first step is made toward the implementation of a solid-state microwave frequency comb.

Additionally, this system is a versatile tool for studying strong nonlinear dynamics. Such an effect is amplitude bistability, which has not been observed in a microwave solid-state spin ensemble coupled to a single-mode cavity so far. This engineered hybrid system approach opens up the possibility for a new route to cavity QED experiments beyond the standard Dicke and Tavis-Cummings model, no only also for truly long-lived quantum memories, solid-state microwave frequency combs and optical-to-microwave quantum transducers.

Acknowledgements

I would like to acknowledge the help and support of my fellow colleagues. I would like to start with my Ph.D. supervisor Johannes Majer. Thank you for your support and trust in me, giving me the ability to pursue experiments beside our main marching direction. Your advice for experimental realizations and knowledge on electronic issues has been vital to me. Also many thanks to Jörg Schmiedmayer who created a terrific working environment at the Atominstitut of the TU Wien. Your enthusiasm for physics and astronomy has been a great inspiration to me. Stefan Rotter and Dmitry Krimer I owe you a debt of gratitude. The collaboration with you has brought my understanding of physics on a new level. An inexpressible amount of thankfulness goes to Bill Munro. Although we are working with a nine hours time shift you have always gave me advice and helped on many subtle issues. Also I would like to acknowledge Michael Trupke, whose knowledge on atoms and cavities has been most helpful to me. Moreover, I have to give a big "thank you" to my Ph.D. examiner Ataç İmamoğlu. I feel deeply honored having you refereeing my work and I am looking forward to receiving your comments.

The work presented in thesis would have not been possible without my dear former Ph.D. colleagues Robert Amsüss, Tobias Nöbauer, and Christian Koller. Their initial work in setting up great parts of the presented experiment has been crucial. The help of Andreas Angerer as my Ph.D. successor on the experiential setup has been vital for the realization of parts presented in this thesis. Together with him I could bring the experiment to a new level and surely he will produce many more beautiful results together with Thomas Astner in the future. I am convinced that their experimental studies will continue to keep a high profile of the laboratory. The work of the master students Abhilash Valookaran and Ralph Glattauer was very helpful during my Ph.D. You might have not noticed, but co-supervising your master theses and watching your fruitful progress was one of most gratifying things during my Ph.D. I have to thank my office colleague Cameron Salter for his proof readings. Also big thanks to Andreas Angerer and Igor Mazets for their patient proofreading. My other fellow Ph.D. colleagues Fritz Diorico and Stefan Minniberger I have to thank for bringing me close to Arnold

Schwarzenegger. All other group members I have to thank for creating a great and joyful work environment, sorry there are too many to list them all. I especially have to thank the doctoral school solids for function of the Austrian science fund for their financial support. Covering my salary at the end of my Ph.D. as well as expenses for equipment was crucial for the work presented here. Also the generous travel reimbursement was important and gave me great opportunities during my Ph.D.

Of course this would all have not been possible without the support of my family and friends. Many thanks to my parents and sister for believing in me and having patience with my career planning. Without your support I would not have come this far. An inexpressible gratitude goes to my love Stephanie Stern. Thank you so much for cheering me up and helping me when ever I needed it the most. I think I have made you proud, at least a bit, and this is also for you!

Contents

Symbols

$\mathcal{H}$	Hamiltonian
ω_s	Central spin eigenfrequency
ω_j	Frequency of the jth Spin
ω_c	Cavity frequency
ω_p	Probe frequency
Δ	Detuning parameter
η	Rate at which power is transmitted into the cavity
Ω	Collective enhanced coupling strength
Ω_R	Rabi frequency
g_j	Single spin cavity coupling strength for the jth spin
g	Homogeneous single cavity spin coupling strength
P_{in}	Power applied at cavity input
$|A|$	Cavity amplitude
$|S_{21}|$	Cavity transmission scattering matrices
Q	Quality factor
ϕ	Phase or angle of the Zeeman field
$|I|, |Q|$	Cavity field quadratures
$\sigma_I, \sigma_Q, \sigma_A$	1σ standard deviation of the quadratures and amplitude
n	Number of measurement shots
$|T|, |R|$	Cavity transmission and reflection
$\mathcal{E}$	Energy
$\mathbf{B}, \vec{B}$	Magnetic cavity field
$\mathbf{E}, \vec{E}$	Electric cavity field
$\in_0$	Vacuum permittivity
$\in_r$	Relative permittivity
μ_0	Vacuum permeability
X	Generalized position operator
P	Generalized momentum operator
x	Position
m	Mass

L	Inductance
C	Capacitance
V	Voltage
I	Current
k	Spring constant or wave vector
c	Speed of light
$|\pm\rangle$	Polariton mode or dressed state
$|B\rangle$	Spin bright or symmetric superradiant state
$|D\rangle$	Spin dark or subradiant states
a and $a^\dagger$	annihilation and creation operator
$\sigma = (\sigma_x, \sigma_y, \sigma_z)$	Spin Pauli matrices
N	Spin or cavity photon number
p_{m_s}	Spin polarization
T_{zf}	Zerofield splitting temperature
D_{zf}, E_{zf}	Zerofield splitting parameters
B_{ext}	Zeeman d.c. external magnetic field
k_B	Boltzmann constant
h and $\hbar$	Planck constant
κ	Cavity dissipation rate
γ	Single spin dissipation rate
γ_e	Electron gyromagnetic ratio
μ_B	Bohr magneton
γ_{hom}	Homogeneous line width
γ_{inh}	Inhomogeneous line width
Γ	Total decoherence rate or with of the polariton modes
Γ_D	Coherence rate of the creates dark states
v	Modulation frequency determining the spectral hole position
$\rho(\omega)$	Spectral spin distribution and density
g_μ	Spectral coupling density
r	Mean spin–spin distance
a_d	Diamond lattice constant
c_d	Diamond atomic density
T_2^{CPMG}	Car-Purcell-Meiboom-Gill Echo recovery time
T_2	Transverse spin lattice relaxation time
T_2^*	Free induction decay time
T_1	Longitudinal spin relaxation time
$T_{1\rho}$	Longitudinal spin lattice relaxation time in the rotating frame
C	Cooperativity
n_0	Saturation photon number

Chapter 1
Introduction and Outline

One of the most basic and important phenomena in our physical world is the interaction of light and matter. Historic examples are the discovery of the spectral light colors by diffraction [1], or the photoelectric effect leading to the insight that light consists of photons [2]. The controlled interaction of atomic systems with electromagnetic radiation was brought to a new regime in recent years [3–7]. This field is known as cavity quantum electro dynamics (cQED) and has been explored in the least few decades and led to a physics Nobel prize in 2012 [8, 9]. The possible coherent exchange of energy between a qubit two-level system and photons have led to many proposed and realized implementations for quantum computation [10–12], simulation [13–15] and communication [16–19] devices and protocols. These techniques herald a new age of technology and could led to a major breakthrough within the 21st century.

On the way to new ground breaking quantum technologies a major landmark was the first demonstration of coherent exchange of single energy quanta in an all solid-state device [20]. Superconducting qubits [21–23] and spin quantum dots [24–27] in circuit QED devices have shown remarkable progress [28–32] since then, and will most probably become a key component in future quantum devices. However, one of their biggest limitations remaining are their short coherence times, limiting the performance especially when compared to atomic systems [33, 34]. A possible way to overcome this technological issue in solid-state devices is to employ electron or nuclear spin impurities in semiconductor crystals [35] which show extremely long coherence times of up to almost one hour [36].

Long-lived spin impurities are prime candidates for the implementation of quantum memories. However, to interface a single defect center is rather difficult, due to their small magnetic moments and weak interaction strengths. This is where the hybridization of spin ensembles becomes important [37–40] in order to coherently exchange energy between a cavity interface and a spin ensemble. High cooperativity and strong coupling has been demonstrated recently, using various different spins systems such as: Nitrogen Vacancy (NV) centers in diamond [41–44], P1 centers

© Springer International Publishing AG 2017

S. Putz, *Circuit Cavity QED with Macroscopic Solid-State Spin Ensembles*,
Springer Theses, DOI 10.1007/978-3-319-66447-7_1

in diamond and ruby ($Al_2O_3:Cr^{3+}$) [45], erbium ($Er^{3+}:Y_2SiO_5$) [46, 47] and phosphorus donors in silicon [48]. Additionally, systems interfacing magnons in a Kittel mode of ferromagnetic yttrium iron garnet (YIG) crystals have shown promising results [49–52].

The coherent storage and retrieval of excitations between spins and a cavity interface has been demonstrated [43, 53–57] in this context already. However, one of the most pressing and unresolved challenges are their short memory times due to inhomogeneous spectral spin broadening [57] induced by the spin host material or by spin-spin interactions. This thesis examines the scaling behavior and possible suppression of decoherence in a hybrid quantum system. The macroscopic solid-state spin ensembles strongly coupled to a single mode cavity gives rise to complex spin dynamics. Therefore to study the effects of inhomogeneous spin broadening in an all solid-state hybrid device is essential. In the presented thesis a device consisting of a superconducting microwave cavity strongly coupled to an ensemble of electron spins hosted by nitrogen-vacancy centers in diamond is realized. This gives the opportunity to understand the underlying effects in these hybrid devices, which is key in the possible implementation of future solid-state quantum memories and simulators.

1.1 Outline

The presented thesis includes a theoretical part in which the principle properties of microwave cavities, and the interaction with macroscopic spin ensembles is discussed. This part is followed by a conceive chapter describing the physical implementation and measurement techniques employed to carry out the presented experiments. The main results presented build on one another and give a detailed insight into:

- **Strong coupling** is demonstrated, in a system in which the single spin interaction strength is much lower than cavity and spin dissipation rates. This is achieved by collective enhancement and allows the direct observation of coherent Rabi oscillations. This chapter lays the foundation of all carried out experiments in the following.
- **Cavity protection effect** the total decoherence in the hybridized system scales with the collective interaction strength. This is due to inhomogeneous spectral spin broadening which is naturally present in macroscopic spin ensembles. However spin dephasing can be suppressed by this effect as will be demonstrated.
- **Spectral hole burning** will be demonstrated and allows the substantial extension of the total coherence times in the system. To my best knowledge this is the first implementation of spectral hole burning in a solid-state hybrid quantum system.
- **Amplitude bistability** will be demonstrated and verify the spin two-level system nature in the coupled hybrid devices. To my best knowledge this is the first demonstration of this effect in a microwave solid-state hybrid quantum system and in general for a largely inhomogeneously broadened spin ensemble.

- **Spin echo spectroscopy** measurements are demonstrated in the experiment with extremely high cooperativity. This techniques allows the estimation of the spin ensemble properties and coherence times. Additionally the limiting mechanisms of the spin coherence times can be understood.

The experimental chapters of this thesis are each designated to a distinct physical effect. Their chronology builds up naturally and immerse deeply into the presented subject of solid-state cavity QED. Although one of the prime goals of such hybrid devices is the development of practical quantum memories, this thesis deals and investigates rather basic but most important aspects of such devices. The presented results demonstrate that hybrid QED devices will also be useful to investigate fundamental physical phenomena in macroscopic spin ensembles in the future. Quantum technologies based on solid-state cavity QED devices promise significant advances in controllable and strongly correlated systems which will be a central research direction in physical sciences in the next decade. It will continue to merge disciplines such as quantum technology, atomic physics, solid-state physics, chemistry and biology. This could allow one to test physical boundaries and to answer some of the biggest questions of our time: What is macroscopicity? Is quantum coherence fundamental or just technological? Answering these issues will change how we understand our physical world and make an important contribution to our society.

References

1. I. Newton, *Opticks, or, a treatise of the reflections, refractions, inflections and colours of light*. William Innys at the West-End of St. Paul's., 1730
2. A. Einstein, Über einen die Erzeugung und Verwandlung des Lichtes betreffenden heuristischen Gesichtspunkt. *Annalen der Physik*, **322**(6):132–148, (Jan 1905). ISSN 1521–3889
3. R.J. Thompson, G. Rempe, H.J. Kimble, Observation of normal-mode splitting for an atom in an optical cavity. Phys. Rev. Lett. **68**(8), 1132–1135 (Feb 1992)
4. G. Rempe, H. Walther, N. Klein, Observation of quantum collapse and revival in a one-atom maser. Phys. Rev. Lett. **58**(4), 353–356 (Jan 1987)
5. A. Furusawa, J.L. Sørensen, S.L. Braunstein, C.A. Fuchs, H.J. Kimble, E.S. Polzik, Unconditional quantum teleportation. Science **282**(5389), 706–709, (Oct 1998). ISSN 0036-8075, 1095–9203
6. J.M. Raimond, M. Brune, S. Haroche, Manipulating quantum entanglement with atoms and photons in a cavity. Rev. Mod. Phys. **73**(3), 565–582 (Aug 2001)
7. D. Leibfried, R. Blatt, C. Monroe, D. Wineland, Quantum dynamics of single trapped ions. Rev. Mod. Phys. **75**(1), 281–324 (Mar 2003)
8. S. Haroche, N. Lecture, Controlling photons in a box and exploring the quantum to classical boundary*. Rev. Mod. Phys. **85**(3), 1083–1102 (July 2013)
9. D.J. Wineland, Nobel lecture: Superposition, entanglement, and raising schrödinger's cat*. Rev. Mod. Phys. **85**(3), 1103–1114 (July 2013)
10. D. Loss, D.P. DiVincenzo, Quantum computation with quantum dots. Phys. Rev. A **57**(1), 120–126 (Jan 1998)
11. J.I. Cirac, P. Zoller, Quantum computations with cold trapped ions. Phys. Rev. Lett. **74**(20), 4091–4094 (May 1995)
12. M.A. Nielsen, I.L. Chuang, Quantum computation and quantum information: 10th anniversary edition. (Cambridge University Press, Dec 2010). ISBN 978-1-139-49548-6

13. R.P. Feynman, Simulating physics with computers. Int. J. Theor. Phys. **21**(6–7), 467–488 (June 1982). ISSN 0020-7748, 1572–9575

14. R. Gerritsma, G. Kirchmair, F. Zähringer, E. Solano, R. Blatt, C.F. Roos, Quantum simulation of the Dirac equation. Nature **463**(7277), 68–71, (Jan 2010). ISSN 0028-0836

15. I. Bloch, J. Dalibard, S. Nascimbène, Quantum simulations with ultracold quantum gases. Nat. Phys. **8**(4), 67–276 (Apr 2012). ISSN 1745–2473

16. D. Bouwmeester, J.-W. Pan, K. Mattle, M. Eibl, H. Weinfurter, A. Zeilinger, Experimental quantum teleportation. Nature **390**(6660), 575–579, (Dec 1997). ISSN 0028-0836

17. L.-M. Duan, M. D. Lukin, J.I. Cirac, P. Zoller. Long-distance quantum communication with atomic ensembles and linear optics. Nature **414**(6862), 413–418 (Nov 2001). ISSN 0028-0836

18. H.J. Kimble. The quantum internet. Nature **453**(7198), 1023–1030, (June 2008). ISSN 0028-0836

19. K. Stannigel, P. Rabl, A.S. Sørensen, P. Zoller, M.D. Lukin, Optomechanical transducers for long-distance quantum communication. Phys. Rev. Lett. **105**(22), 220501 (Nov 2010)

20. Y. Nakamura, Yu A. Pashkin, J.S. Tsai, Coherent control of macroscopic quantum states in a single-Cooper-pair box. Nature **398**(6730), 786–788 (April 1999). ISSN 0028-0836

21. A. Wallraff, D.I. Schuster, A. Blais, L. Frunzio, R.-S. Huang, J. Majer, S. Kumar, S.M. Girvin, R.J. Schoelkopf, Strong coupling of a single photon to a superconducting qubit using circuit quantum electrodynamics. Nature **431**, 162–167 (Sept 2004)

22. A.A. Houck, D.I. Schuster, J.M. Gambetta, J.A. Schreier, B.R. Johnson, J.M. Chow, L. Frunzio, J. Majer, M.H. Devoret, S.M. Girvin, R. J. Schoelkopf, Generating single microwave photons in a circuit. Nature, **449**(7160), 328–331, (Sept 2007). ISSN 0028-0836

23. T. Niemczyk, F. Deppe, H. Huebl, E. P. Menzel, F. Hocke, M. J. Schwarz, J. J. Garcia-Ripoll, D. Zueco, T. Hümmer, E. Solano, A. Marx, R. Gross. Circuit quantum electrodynamics in the ultrastrong-coupling regime. Nat. Phys. **6**(10), 772–776, (Oct 2010). ISSN 1745-2473

24. R. Hanson, L.P. Kouwenhoven, J.R. Petta, S. Tarucha, L.M.K. Vandersypen, Spins in few-electron quantum dots. Rev. Mod. Phys. **79**(4), 1217–1265 (Oct 2007)

25. J.R. Petta, A.C. Johnson, J.M. Taylor, E.A. Laird, A. Yacoby, M.D. Lukin, C.M. Marcus, M.P. Hanson, A.C. Gossard, Coherent manipulation of coupled electron spins in semiconductor quantum dots. Science **309**(5744), 2180–2184, (Sept 2005). ISSN 0036-8075, 1095–9203

26. J.J. Viennot, M.C. Dartiailh, A. Cottet, T. Kontos, Coherent coupling of a single spin to microwave cavity photons. Science **349**(6246), 408–411, (July 2015). ISSN 0036-8075, 1095–9203

27. Y.-Y. Liu, J. Stehlik, C. Eichler, M.J. Gullans, J.M. Taylor, J.R. Petta, Semiconductor double quantum dot micromaser. Science **347**(6219), 285–287, (Jan 2015). ISSN 0036-8075, 1095–9203

28. R.J. Schoelkopf, S.M. Girvin, Wiring up quantum systems. Nature 451(7179), 664–669, (Feb 2008). ISSN 0028-0836

29. M. Hofheinz, H. Wang, M. Ansmann, R.C. Bialczak, E. Lucero, M. Neeley, A.D. O'Connell, D. Sank, J. Wenner, J.M. Martinis, A.N. Cleland, Synthesizing arbitrary quantum states in a superconducting resonator. Nature **459**(7246), 546–549, (May 2009). ISSN 0028-0836

30. L. DiCarlo, J.M. Chow, J.M. Gambetta, L.S. Bishop, B.R. Johnson, D.I. Schuster, J. Majer, A. Blais, L. Frunzio, S.M. Girvin, R.J. Schoelkopf, Demonstration of two-qubit algorithms with a superconducting quantum processor. Nature **460**(7252), 240–244, (July 2009). ISSN 0028-0836

31. K.D. Petersson, L.W. McFaul, M.D. Schroer, M. Jung, J.M. Taylor, A.A. Houck, J.R. Petta, Circuit quantum electrodynamics with a spin qubit. Nature **490**(7420), 380–383, (Oct 2012). ISSN 0028-0836

32. J. Kelly, R. Barends, A.G. Fowler, A. Megrant, E. Jeffrey, T.C. White, D.Sank, J.Y. Mutus, B. Campbell, Y. Chen, Z. Chen, B. Chiaro, A. Dunsworth, I.-C. Hoi, C. Neill, P.J.J. O'Malley, C. Quintana, P. Roushan, A. Vainsencher, J. Wenner, A.N. Cleland, J. M. Martinis, State preservation by repetitive error detection in a superconducting quantum circuit. Nature **519**(7541), 66–69, (Mar 2015). ISSN 0028-0836

33. P. Treutlein, P. Hommelhoff, T. Steinmetz, T.W. Hänsch, J. Reichel, Coherence in microchip traps. Phys. Rev. Lett. **92**(20), 203005 (May 2004)
34. C. Langer, R. Ozeri, J.D. Jost, J. Chiaverini, B. DeMarco, A. Ben-Kish, R.B. Blakestad, J. Britton, D.B. Hume, W.M. Itano, D. Leibfried, R. Reichle, T. Rosenband, T. Schaetz, P.O. Schmidt, D.J. Wineland, Long-lived qubit memory using atomic ions. Phys. Rev. Lett. **95**(6), 060502 (Aug 2005)
35. R. Hanson, D.D. Awschalom, Coherent manipulation of single spins in semiconductors. Nature **453**(7198), 1043–1049, (June 2008). ISSN 0028-0836
36. K. Saeedi, S. Simmons, J.Z. Salvail, P. Dluhy, H. Riemann, N.V. Abrosimov, P. Becker, H.-J. Pohl, J.J.L. Morton, M.L.W. Thewalt, Room-temperature quantum bit storage exceeding 39 minutes using ionized donors in silicon-28. Science **342**(6160), 830–833, (Nov 2013). ISSN 0036-8075, 1095-9203
37. J. Verdú, H. Zoubi, Ch. Koller, J. Majer, H. Ritsch, J. Schmiedmayer, Strong magnetic coupling of an ultracold gas to a superconducting waveguide cavity. Phys. Rev. Lett. **103**(4), 043603 (July 2009)
38. A. İmamoğlu, Cavity QED based on collective magnetic dipole coupling: Spin ensembles as hybrid two-level systems. Phys. Rev. Lett. **102**(8), 083602 (Feb 2009)
39. Z.-L. Xiang, S. Ashhab, J.Q. You, F. Nori, Hybrid quantum circuits: Superconducting circuits interacting with other quantum systems. Rev. Mod. Phys. **85**(2), 623–653 (April 2013)
40. G. Kurizki, P. Bertet, Y. Kubo, K. Mølmer, D. Petrosyan, P. Rabl, J. Schmiedmayer, Quantum technologies with hybrid systems. *Proceedings of the National Academy of Sciences*, **112**(13), 3866–3873, (Mar 2015). ISSN 0027-8424, 1091–6490
41. Y. Kubo, F.R. Ong, P. Bertet, D. Vion, V. Jacques, D. Zheng, A. Dréau, J.-F. Roch, A. Auffeves, F. Jelezko, J. Wrachtrup, M.F. Barthe, P. Bergonzo, D. Esteve, Strong coupling of a spin ensemble to a superconducting resonator. Phys. Rev. Lett. **105**(14), 140502 (Sept 2010)
42. R. Amsüss, Ch. Koller, T. Nöbauer, S. Putz, S. Rotter, K. Sandner, S. Schneider, M. Schramböck, G. Steinhauser, H. Ritsch, J. Schmiedmayer, J. Majer, Cavity QED with magnetically coupled collective spin states. Phys. Rev. Lett. **107**(6), 060502 (Aug 2011)
43. X. Zhu, S. Saito, A. Kemp, K. Kakuyanagi, S. Karimoto, H. Nakano, W.J. Munro, Y. Tokura, M. Everitt, K.S. Nemoto, M. Kasu, N. Mizuochi, K. Semba, Coherent coupling of a superconducting flux qubit to an electron spin ensemble in diamond. Nature **478**, 221–224 (Sept 2011)
44. C. Grèzes, *Towards a spin ensemble qunatum memory for superconducting qubits*. PhD Thesis, Docteur de l'Université Pierre et Marie Curie, CEA Saclay, 2015
45. D.I. Schuster, A.P. Sears, E. Ginossar, L. DiCarlo, L. Frunzio, J.J.L. Morton, H. Wu, G.A.D. Briggs, B.B. Buckley, D.D. Awschalom, R.J. Schoelkopf, High-cooperativity coupling of electron-spin ensembles to superconducting cavities. Phys. Rev. Lett. **105**(14), 140501 (Sept 2010)
46. S. Probst, H. Rotzinger, S. Wünsch, P. Jung, M. Jerger, M. Siegel, A.V. Ustinov, P.A. Bushev, Anisotropic rare-earth spin ensemble strongly coupled to a superconducting resonator. Phys. Rev. Lett. **110**(15), 157001 (April 2013)
47. S. Probst, *Hybrid quantum system based on rare earth doped crystals*. PhD Thesis, Karlsruher Instituts für Technologie (KIT), Karlsruhe, 2015
48. C.W. Zollitsch, K. Mueller, D.P. Franke, S.T.B. Goennenwein, M.S. Brandt, R. Gross, H. Huebl, High cooperativity coupling between a phosphorus donor spin ensemble and a superconducting microwave resonator. Appl. Phys. Lett. **107**(14), 142105, (Oct 2015). ISSN 0003-6951, 1077–3118
49. H. Huebl, C.W. Zollitsch, J. Lotze, F. Hocke, M. Greifenstein, A. Marx, R. Gross, T.B. Sebastian, Goennenwein, High cooperativity in coupled microwave resonator ferrimagnetic insulator hybrids. Phys. Rev. Lett. **111**(12), 127003 (Sept 2013)
50. Y. Tabuchi, S. Ishino, T. Ishikawa, R. Yamazaki, K. Usami, Y. Nakamura, Hybridizing ferromagnetic magnons and microwave photons in the quantum limit. Phys. Rev. Lett. **113**(8), 083603 (Aug 2014)

51. M. Goryachev, W.G. Farr, D.L. Creedon, Y. Fan, M. Kostylev, M.E. Tobar, High-cooperativity cavity QED with magnons at microwave frequencies. Phys. Rev. Appl. **2**(5), 054002 (2014)
52. Y. Tabuchi, S. Ishino, A. Noguchi, T. Ishikawa, R. Yamazaki, K. Usami, Y. Nakamura, Coherent coupling between a ferromagnetic magnon and a superconducting qubit. Science **349**(6246), 405–408, (July 2015). ISSN 0036-8075, 1095–9203
53. W. Hua, R.E. George, J.H. Wesenberg, K. Mølmer, D.I. Schuster, R.J. Schoelkopf, K.M. Itoh, A. Ardavan, J.J.L. Morton, G.A.D. Briggs, Storage of multiple coherent microwave excitations in an electron spin ensemble. Phys. Rev. Lett. **105**(14), 140503 (Sept 2010)
54. Y. Kubo, C. Grèzes, A. Dewes, T. Umeda, J. Isoya, H. Sumiya, N. Morishita, H. Abe, S. Onoda, T. Ohshima, V. Jacques, A. Dréau, J.-F. Roch, I. Diniz, A. Auffeves, D. Vion, D. Esteve, P. Bertet, Hybrid quantum circuit with a superconducting qubit coupled to a spin ensemble. Phys. Rev. Lett. **107**(22), 220501 (Nov 2011)
55. Y. Kubo, I. Diniz, A. Dewes, V. Jacques, A. Dréau, J.-F. Roch, A. Auffeves, D. Vion, D. Esteve, P. Bertet, Storage and retrieval of a microwave field in a spin ensemble. Phys. Rev. A **85**(1), 012333 (Jan 2012)
56. C. Grèzes, B. Julsgaard, Y. Kubo, M. Stern, T. Umeda, J. Isoya, H. Sumiya, H. Abe, S. Onoda, T. Ohshima, V. Jacques, J. Esteve, D. Vion, D. Esteve, K. Mølmer, Bertet P. Multi-mode storage and retrieval of microwave fields in a spin ensemble, http://arxiv.org/abs/1401.7939, 2014
57. S. Putz, D.O. Krimer, R. Amsüss, A. Valookaran, T.Nöbauer, J. Schmiedmayer, S. Rotter, J. Majer, Protecting a spin ensemble against decoherence in the strong-coupling regime of cavity QED. Nat. Phys. **10**(10), 720–724, (Oct 2014). ISSN 1745-2473

Chapter 2
Confined Electromagnetic Waves—Cavities

In this chapter I will briefly focus on the description of the electromagnetic field inside a cavity. Starting from Maxwell's equations, one can derive a formalism treating such field modes inside a resonator as harmonic oscillators. These field modes can then be quantized and treated as quantum harmonic oscillators. In the latter, I will use this formalism to describe electrical circuits and microwave cavities. This concepts are a broad field for itself and are a key in quantum optics and cavity quantum electro dynamics (cQED), and a comprehensive introduction can be found in the following textbooks [1–8] on which the following sections are based on.

2.1 Electromagnetic Radiation

2.1.1 Mode Expansion in Free Space

The starting point for describing the electromagnetic field in free space is the set of Maxwell equations

$$\vec{\nabla}\vec{E} = 0, \ \vec{\nabla}\vec{B} = 0, \ \vec{\nabla} \times \vec{E} = -\frac{\partial}{\partial t}\vec{B} \text{ and } \vec{\nabla} \times \vec{B} = \frac{1}{c^2}\frac{\partial}{\partial t}\vec{E} \qquad (2.1)$$

assuming no free charges and currents. Derived from Eq. 2.1, wave equations for the electric $\vec{E}$ and magnetic $\vec{B}$ field components read

$$\Delta\vec{E} - \frac{1}{c^2}\frac{\partial^2}{\partial^2}\vec{E} = 0 \text{ and } \Delta\vec{B} - \frac{1}{c^2}\frac{\partial^2}{\partial^2}\vec{B} = 0 . \qquad (2.2)$$

For which a solution can be found using a wave expansion for electro magnetic field components

© Springer International Publishing AG 2017
S. Putz, *Circuit Cavity QED with Macroscopic Solid-State Spin Ensembles*,
Springer Theses, DOI 10.1007/978-3-319-66447-7_2

Fig. 2.1 Visualization of a
transverse electromagnetic
(TEM) wave

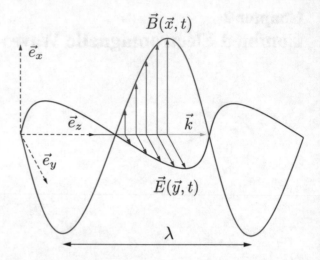

$$\vec{E}_y = \mathrm{i}\frac{\omega}{c\sqrt{V}} \sum_k \vec{e}_y A(t) \mathrm{e}^{\mathrm{i}k\vec{z}} + c.c.\,\mathrm{and}$$

$$\vec{B}_x = \mathrm{i}\frac{1}{\sqrt{V}} \sum_k \vec{k} \times \vec{e}_y A(t) \mathrm{e}^{\mathrm{i}k\vec{z}} + c.c. \qquad (2.3)$$

with $A(t) = A_0 \mathrm{e}^{-\mathrm{i}\omega t}$. It follows that the amplitudes of the electric $\vec{E}$ and magnetic field $\vec{B}$ are in phase but perpendicular to each other. The polarization of the magnetic field $\vec{e}_x$ follows immediately by choosing the polarization direction of the electric field in the $\vec{e}_y$ direction and letting the wave propagate in the $\vec{e}_z$ direction, denoted by the wave vector $\vec{k} = \vec{e}_x \times \vec{e}_y$. Such purely transversal polarized electromagnetic waves are known as transverse electromagnetic (TEM) waves (Fig. 2.1). In free space there is an infinite number of possible wave modes with different wavelengths λ and frequencies following the dispersion relation $\omega = kc$ and therefore one has to sum over all possible $k = |\vec{k}|$ values in Eq. 2.3.

2.1.2 Modes Inside a Cavity

In the previous section the electromagnetic field in free space was described by a mode expansion. Thus the energy of the electromagnetic field inside a volume V follows by summation or integration over a continuum of all possible field modes. The number of modes in a volume V can be restricted by defining boundary conditions for the electromagnetic field. Here boundary condition simply mean that the electromagnetic field has to obey constraints on interfaces like walls. This is realized by assuming e.g. two reflecting surfaces spatially separated by a distance l. On the mirrors incident $\vec{E}_i$ and reflected electric field $\vec{E}_r$ experience a π phase shift, since $\vec{k} = \vec{E}_i \times \vec{B}_i =$

$-\vec{E}_r \times \vec{B}_r$ must be fulfilled. Reflections and phase shifts on both surfaces create interference and standing waves between the mirrors are formed. For such a standing wave, maxima and minima of the electric and magnetic field amplitudes do have a $\pi/2$ phase shift. Therefore after performing a wave expansion for the field inside the cavity the electromagnetic field components read

$$E_y(z,t) = \frac{1}{\sqrt{V}} \sum_n \frac{\omega_c}{c} \sin(k_n z) A_0 (e^{i\omega_c t} + e^{-i\omega_c t})$$

$$B_x(z,t) = \frac{i}{\sqrt{V}} \sum_n \frac{\omega_c}{c} \cos(k_n z) A_0 (e^{i\omega_c t} - e^{-i\omega_c t}) \qquad (2.4)$$

with $k_n = \pi n/l$. Therefore the condition for creating a standing wave between the two mirrors (Fig. 2.2) (i.e. resonance) is given by the distance l which must be an integer multiple n of half of the wavelength i.e. $\lambda/2$ and $k_n = \frac{2\pi n}{\lambda}$. This means that after one round trip, the phase shift is equal to the initial phase of the wave. Hence a resonance condition follows and gives a fundamental resonance of the cavity field mode at $\omega_c = \frac{2\pi}{\lambda} c$.

The derived expressions for the electric and magnetic field inside the resonator, stated in Eq. 2.4, would include all higher harmonics of the cavity. However, in the latter only the fundamental resonance $\omega_c = \pi/l$ is of interest and all higher harmonics are ignored. This is usally assumed in cavity QED and known as a single mode cavity. The total energy in the cavity and single field mode ($n = 1$) is then given by integrating over the entire cavity volume V

$$\mathcal{E} = \frac{1}{2} \int_V (\epsilon_0 E_x^2 + \frac{1}{\mu_0} B_y^2) dV \rightarrow \frac{1}{2} \frac{\omega^2}{c^2} A_0^2 \qquad (2.5)$$

with vacuum permittivity ϵ_0 permeability μ_0. The total energy is thus proportional to the squared amplitude of $A(t)$ introduced in the previous section. The total energy $\mathcal{E}$ is constant but oscillating between electric and magnetic field components with the cavity eigenfrequency ω_c. This fact allows to treat the electric and magnetic field

Fig. 2.2 Visualization of a the electromagnetic field between two mirrors creating a standing wave in a $\lambda/2$ cavity. The stored energy is oscillating between magnetic and electric field amplitudes with the cavity frequency ω_c

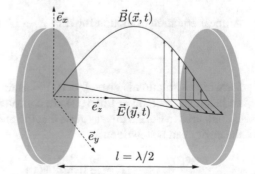

amplitudes inside the cavity as real and complex parts or field quadratures of the electromagnetic field as $A = E_x + iB_y$. As we already know E_x and B_y experience a $\pi/2$ phase shift inside a cavity and span what we will later call phase space.

2.2   Single Cavity Modes—Harmonic Oscillators

2.2.1   Canonical Variables of the Electromagnetic Field

In the previous section (Eq. 2.5) the total energy of the electromagnetic field inside a cavity was derived. Here we show how a single field mode can be treated as a single harmonic oscillator. If the resonator is restricted to the fundamental resonance (i.e. single mode), the energy density is given by

$$\mathcal{H} = \frac{1}{2}(\epsilon_0 E^2 + \frac{1}{\mu_0} B^2) \tag{2.6}$$

with vacuum permittivity ϵ_0 permeability μ_0 and is a Hamiltonian function $\mathcal{H}$. As already discussed E and B are the field quadratures and are $\pi/2$ phase shifted, which is basically a set of canonical variables. The energy of the electromagnetic wave inside a cavity can be interpreted to be equivalent to the sum of the kinetic and potential energy of the system and allows to map this problem to the harmonic oscillator model from classical mechanics. In such, the total system energy can be written as a Hamiltonian reading

$$\mathcal{H} = \frac{1}{2}(\frac{p^2}{2m} + kx^2) \tag{2.7}$$

with momentum p, position x and force constant k. Both x and p can be renormalized such that

$$\mathcal{H} = \frac{\omega}{2}(X^2 + P^2) . \tag{2.8}$$

A linear equation of motion for $A = X + iP$ follows

$$\omega^2 A + \ddot{A} = 0 \tag{2.9}$$

for which a solution is given by $A(t) = A_0 e^{-i\omega t}$. Hence position $X(t) = \text{Re}(A(t))$ and momentum $P(t) = \text{Im}(A(t))$ are complex and imaginary parts of $A = X + iP$. If compared to Eq. 2.5 this suggests that the energy of the electromagnetic field inside the cavity can be rewritten as

$$\mathcal{E} = \frac{\omega}{2}\{|A_0 \cos(\omega t)|^2 + |A_0 \sin(\omega t)|^2\} \tag{2.10}$$

with a field amplitude $|A| = \sqrt{\mathrm{Re}(A)^2 + \mathrm{Im}(A)^2}$ and phase $\phi = \tan\frac{\mathrm{Re}(A)}{\mathrm{Im}(A)}$ for the cavity field. Note that in comparison to the previous section $A(t)$ is renormalized and directly related to X and P by $\omega P = \dot{X}$. Form Hamilton's equations $\frac{\partial \mathcal{H}}{\partial X} = -\dot{P}$ and $\frac{\partial \mathcal{H}}{\partial P} = \dot{X}$ one further can conclude that X and P are canonical variables.

2.2.2 Drive and Dissipation of a Classical Oscillator

As shown above the electromagnetic field inside a cavity can be described by a harmonic oscillator. However, before quantizing the resonator field mode it is instructive to look on a classical damped and driven harmonic oscillator in order to give an intuitive picture first (Fig. 2.3).

The equation of motion for a mass m on a spring k_i with an internal damping constant γ_i reads

$$\ddot{x} + 2\left(\frac{\gamma_i + \gamma_c}{m}\right)\dot{x} + \left(\frac{k_i + k_c}{m}\right)x = \frac{k_c}{m}De^{i\omega t} \tag{2.11}$$

with position x off the mass. A driving force is applied by a damped γ_c spring k_c which is periodically deflected with an amplitude D and frequency ω. By defining a frequency $\sqrt{\frac{k_c+k_i}{m}} = \omega_0$ the equation of motion can be transformed to

$$\ddot{x} + 2\gamma\dot{x} + \omega_0^2 x = \omega^2 De^{i\omega t} \tag{2.12}$$

with $\gamma = (\gamma_i + \gamma_c)/m$. On should note that any coupling of a harmonic oscillator to an additional spring also changes its true eigenfrequency $\sqrt{\frac{k_i}{m}}$. With the Ansatz $x = A_0 e^{i\omega t}$ a solution (i.e. $\dot{A} = 0$) for the stationery amplitude

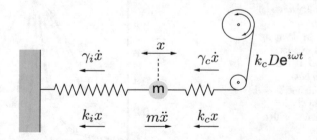

Fig. 2.3 Schematic drawing of a classical harmonic oscillator. A mass m on a spring k_i determines the natural eigenfrequency of the damped oscillator. Coupled to a second damped spring k_c this can be interpreted as a one sided optical cavity

$$|A_0|^2 = \frac{\omega^4 D^2}{(\omega_0^2 - \omega^2)^2 + (\omega 2\gamma)^2} \tag{2.13}$$

is found. The amplitude on resonance (i.e. $\omega_0^2 - \omega^2 = 0$) can be derived as

$$|A_{\text{Res}}| = \frac{\omega}{2\gamma}D = QD \tag{2.14}$$

with the introduced quantity $Q = \omega/2\gamma$. The resonator quality factor Q states that the resonator amplitude is equal to the driving amplitude enhanced by a factor of Q. One should note that in the latter the quality factor Q is defined in the standard way by the ratio of stored to lost energy rather than terms of resonator amplitudes.

For cavities with high Q factors Eq. 2.13 can be well approximated by

$$|A|^2 = \frac{\eta^2}{(\omega_0 - \omega)^2 + \kappa^2} \tag{2.15}$$

if the damping constant γ becomes small. This Lorentzian function with a full-width at half-maximum (FWHM) line width 2κ and drive amplitude $\eta = \sqrt{\omega \kappa}D$ will be used in the latter to describe the cavity response. In Fig. 2.4 the difference between Eqs. 2.13 and 2.15 is plotted and also shows how the line shape of a harmonic oscillator approaches a Lorentzian function for small values of γ. From the Ansatz $x = Ae^{i\omega t}$ it follows that the amplitude A oscillates sinusoidaly with frequency ω, thus a constant energy transfer between kinetic and potential stored energy. After the periodic driving force is switched off ($D = 0$) the amplitude $|A(t)|$ will decay with the rate κ while the intensity $|A(t)|^2$ decays twice as fast with 2κ. One should note that the squared amplitude of the harmonic oscillator has to be compared to the Lorentz function in order to estimate κ.

Fig. 2.4 Difference between Lorentzian oscillator model and low Q harmonic oscillators. As the damping constant γ gets smaller (*yellow* $\gamma = \omega_0/10$, *red* $\gamma = \omega_0/20$ and *blue* $\gamma = \omega_0/100$) the line is well approximated by a Lorentzian function with a line width of 2κ

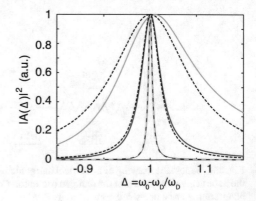

2.2.3 Quantum Harmonic Oscillator

A single cavity mode can be treated as a harmonic oscillator and quantized in full quantum mechanical fashion. The starting point is the canonical transformation of the cavity field $A = X + iP$ to the quantum mechanical creation and annihilation operators $\hat{a}$ and $\hat{a}^\dagger$ reading

$$\hat{a} = \frac{\hat{X} + i\hat{P}}{\sqrt{2\hbar}} \text{ and } \hat{a}^\dagger = \frac{\hat{X} - i\hat{P}}{\sqrt{2\hbar}} . \tag{2.16}$$

With the commutation relation for position and momentum operator $[\hat{P}, \hat{X}] = -i\hbar$ the quantized expression for the Hamiltonian for a single mode cavity follows as

$$\mathcal{H} = \hbar\omega(a^\dagger a + \frac{1}{2}) \tag{2.17}$$

where the operator hats in $\hat{a}(\hat{a}^\dagger) \rightarrow a(a^\dagger)$ were dropped as in the rest of this thesis.

The mean energy stored in the system is equal to the number of stored excitations and an operator for the photon number then reads $N = a^\dagger a$. This operator uses a new eigenbasis representation for the energy in the cavity called Fock space or number states $\rangle n$. Creation and annihilation operators $a^\dagger$ and a raise and lower the number of excitations, since $a^\dagger|n\rangle = \sqrt{n+1}|n+1\rangle$ and $a|n\rangle = \sqrt{n}|n-1\rangle$. This also means that although the cavity is empty $|0\rangle$ there will be a zero point energy $\hbar\omega/2$ due to vacuum fluctuations. Since the Hamiltonian is an expression combining kinetic and potential energy we can calculate its eigenstates or wavefunction by solving the time independent Schrödinger equation

$$\mathcal{H}\rangle\Psi = \mathcal{E}\rangle\Psi \tag{2.18}$$

where $\mathcal{E}$ is an energy eigenvalues found by Eq. 2.17. One finds a set of solutions for $|\Psi(x)\rangle$ using the Hermite polynomials [9] and can calculate the probability $|\Psi(x)|^2$ of finding the oscillator in the potential given by $\frac{\omega^2}{2}X^2$. As is shown in Fig. 2.5 by the plotted squared wave function of the quantum harmonic oscillator, the energy variance in the system becomes more and more classical in a sense when the number of excitations grows.

The quantized harmonic oscillator model and its ladder operators $a^\dagger$ and a, can be used to quantize the electromagnetic field inside the cavity. The field quadratures read

$$X = \text{Re}(A) = \sqrt{\frac{\hbar}{2}}(a + a^\dagger) \rightarrow E_x = \sqrt{\frac{\hbar\omega}{2\epsilon_0 V}}(a + a^\dagger)\sin(kz)$$

$$P = \text{Im}(A) = -i\sqrt{\frac{\hbar}{2}}(a - a^\dagger) \rightarrow B_y = -i\sqrt{\frac{\hbar\omega\mu_0}{2V}}(a - a^\dagger)\cos(kz) \tag{2.19}$$

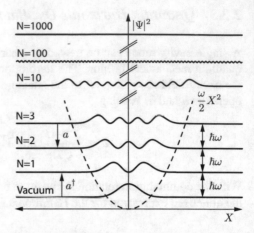

Fig. 2.5 The probability distribution of the quantum harmonic oscillator $|\Psi(x)|^2$ of finding the oscillator somewhere in the potential given by $\frac{\omega^2}{2}X^2$. The wave function and hence the probability of the oscillator becomes more and more de-localized for growing number of excitations $\langle N \rangle$

in a unit volume of size V with the vacuum permittivity ϵ_0 and permeability μ_0. The electromagnetic field is therefore quantized in the same way as the quantum harmonic oscillator. From Eq. 2.19 one finds that the expectation values for the electromagnetic field components read

$$\langle n|E|n \rangle = 0 = \langle n|B|n \rangle \tag{2.20}$$

and become zero. For the described harmonic oscillator this is equivalent to having zero amplitude, which is a possible but rather special case. Therefore, a formalism to introduce states with well defined field amplitudes are coherent states

$$a|\alpha \rangle = \alpha|\alpha \rangle \tag{2.21}$$

which are eigenstates of the annihilation operator a. These states $\alpha = |\alpha|e^{i\phi}$, with amplitude $|\alpha|$ and phase ϕ, can be expanded in the Fock or number basis states

$$|\alpha \rangle = e^{-\frac{1}{2}|\alpha|^2} \sum_n \frac{\alpha^n}{\sqrt{n!}} |n \rangle \tag{2.22}$$

since coherent states $\alpha|\alpha \rangle$ are eigenstates of a. The probability of detecting n photons in a coherent state $|\langle n|\alpha \rangle|^2$ obeys a Poisonian distribution and becomes Gaussian for large $|\alpha|$. The expectation values of the field quadratures therefore read

$$\langle X \rangle = \sqrt{\frac{\hbar}{2}}(\alpha + \alpha^*) = \text{Re}(\alpha)$$

$$\langle P \rangle = -i\sqrt{\frac{\hbar}{2}}(\alpha - \alpha^*) = \text{Im}(\alpha) \tag{2.23}$$

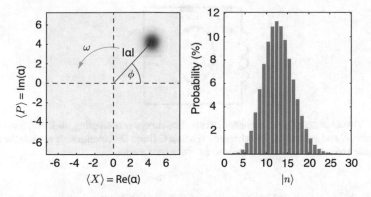

Fig. 2.6 *Left* Visualization of a coherent cavity state with amplitude $|\alpha| = \sqrt{\langle X \rangle^2 + \langle P \rangle^2}$ rotating with ω in phase space. *Right* Since coherent states are eigenstates of the annihilation operator a they can be expanded in the Fock basis $|n\rangle$. For a state $|\alpha\rangle$ we find a photon distribution $\langle N \rangle = |\alpha|^2$ with variance $\Delta N = |\alpha|$

and one finds the average energy of a coherent cavity state to be

$$\langle \mathcal{E} \rangle = \hbar\omega |\alpha|^2 = \hbar\omega \langle N \rangle \tag{2.24}$$

with variance $\Delta \mathcal{E} = \hbar\omega |\alpha|$. For large amplitudes the relative variance $\Delta \mathcal{E} / \langle \mathcal{E} \rangle$ goes to zero and the state becomes more and more well defined or "classical" as is shown in Fig. 2.6.

2.3 Electrical Oscillators—"From Classical to Quantum"

2.3.1 Quantization of a LC Oscillator

In this section an electric LC circuit is discussed as a harmonic oscillator which can be fully quantized. The previously introduced formalism is used to describe such a circuit which can be considered as the counterpart to an optical cavity. The starting point as is shown in Fig. 2.7 is a circuit consisting of an inductance L and capacitor C in a parallel configuration. The inductance is a measure of how much magnetic flux Φ is created in a circuit carrying a given current I, hence the relation $L = \Phi / I$. Additionally it follows from Faraday's law that by an an alternating current a voltage $V_L = \frac{d\Phi}{dt} = L\frac{dI}{dt}$ is induced in the circuit. Contrary, the capacitance C of two parallel plates is given by the ratio of the charging[1] $\pm Q$ and the voltage V across both plates, given by $V_C = Q/C$. For a closed circuit the potential across the circuit has to be

---

[1]*Nota bene:* in this section the charge is abbreviated by capital Q and should not be confused with the cavity quality factor Q through out the rest of this thesis.

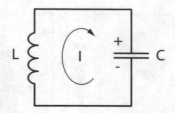

Fig. 2.7 Visualization of a LC resonant circuit. The energy is oscillating back and forth between the capacitor C and inductor L. The charged capacitor C stores the kinetic energy, while the inductor L stores the kinetic energy of the harmonic LC oscillator model

zero ($V_C + V_L = 0$) and at any node of the circuit the sum of in and out flowing currents hast to be zero ($I_C = I_L$) according to Kirchhoff's circuit laws. From this the following equation of motion for an electrical LC circuit

$$Q + \frac{1}{LC}\ddot{Q} = 0 \tag{2.25}$$

can be derived with a resonance frequency given by $\omega = \frac{1}{\sqrt{LC}}$ and characteristic impedance $Z = \sqrt{\frac{L}{C}}$. The energy stored in the inductance L and capacitance C is given by $\mathcal{E}_L = \frac{1}{2}LI^2 = \frac{\Phi^2}{2L}$ and $\mathcal{E}_C = \frac{1}{2}CV^2 = \frac{Q^2}{2C}$, respectively.

With the total energy density in the circuit one can make a transition to Hamilton's mechanic

$$\mathcal{H} = \frac{1}{2}\left(\frac{Q^2}{C} + L\dot{Q}^2\right) = \frac{1}{2}\left(\frac{Q^2}{C} + \frac{\Phi^2}{L}\right) \tag{2.26}$$

with $\partial\mathcal{H}/\partial\Phi = \dot{Q} = \Phi/L$ and $\partial\mathcal{H}/\partial Q = -\dot{\Phi} = Q/C$. The magnetic flux Φ is considered as the generalized momentum with kinetic energy $\mathcal{E}_{\text{kin}} = \frac{\Phi^2}{2L}$ stored in the inductor. Whereas the electric charge Q is the generalized coordinate connected to the potential energy stored in the capacitor $\mathcal{E}_{\text{pot}} = \frac{Q^2}{2C}$. Therefore the electric charge and magnetic flux are the canonical variables of the harmonic oscillator. This circuit can be quantized in a quantum mechanical way by using the commutator relation $[\Phi, Q] = -\mathrm{i}\hbar$. Annihilation a and creation $a^\dagger$ operators can be expressed in terms of Q and Φ as

$$Q = \sqrt{\frac{\hbar}{2Z}}(a + a^\dagger) \text{ and } \Phi = -\mathrm{i}\sqrt{\frac{\hbar Z}{2}}(a - a^\dagger) \tag{2.27}$$

and the Hamiltonian for a loss-less quantized LC circuit reads

$$\mathcal{H} = \hbar\omega\left(a^\dagger a + \frac{1}{2}\right). \tag{2.28}$$

In analogy to the quantization of the electromagnetic field inside a cavity, voltage and current operator reading

$$V = \sqrt{\frac{\hbar\omega}{2C}}(a + a^{\dagger}) \text{ and } I = -i\sqrt{\frac{\hbar\omega}{2L}}(a - a^{\dagger}) \tag{2.29}$$

can be found as field quadratures. The zero point energy in the cavity, $(a^{\dagger}a = 0)$, caused by vacuum fluctuations is then given by

$$\mathcal{E}_{\text{vac}} = \mathcal{E}_C + \mathcal{E}_L = \frac{1}{2}(CV^2 + LI^2) = \frac{\hbar\omega}{2} . \tag{2.30}$$

2.3.2 Drive and Dissipation in a Classical LC Oscillator

After deriving an equation of motion and the quantization of a simple electrical LC resonator dissipation and coupling to external ports has to be added to such a circuit. In Fig. 2.8 a circuit diagram of a lumped element resonator in parallel LC configuration is shown. The circuit includes an internal resistance R accounting for dissipation and is coupled by two capacitors C_k to a feedline with $Z_0 = 50\,\Omega$ at port one and two. The input impedance of the uncoupled classical electrical resonator is given by [8]

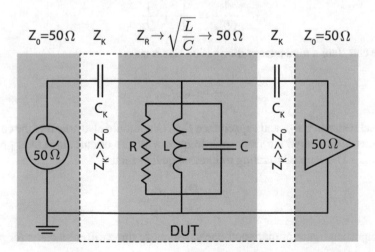

Fig. 2.8 Lumped element equivalent circuit diagram of a RLC resonant circuit connected to a $50\,\Omega$ load via two coupling capacitors C_k. The total quality-factor Q_{tot} has contributions from the internal quality factor Q_I determined by R and a coupling quality-factor Q_k which is $\propto C_k^2$

$$Z_{in} = \left(\frac{1}{R} + \frac{1}{j\omega L} + j\omega C \right)^{-1} \tag{2.31}$$

with a characteristic impedance of the circuit $Z_R = \sqrt{\frac{L}{C}}$. The energy dissipated by the resonator is given by $P_d = \frac{1}{2}|V|^2/R$ and the maximum energy stored is given by $\mathcal{E} = \frac{1}{2}V^2 C$ since on resonance electric and magnetic energy are equal. For such a parallel RLC resonator one can then derive an quantity for the internal quality factor which reads

$$Q_I = \omega RC = \frac{\omega}{\kappa_I} \tag{2.32}$$

and is the relation of maximum stored to dissipated energy per cycle. The quality factor depends on the internal resistance which determines the total dissipation. By defining $R = Z_0/\alpha$ with an attenuation α the Q factor would go to infinity for a perfect loss less LC resonator.

If a feedline with characteristic impedance $Z_0 = 50\ \Omega$ and a $50\ \Omega$ load R_L is coupled to the cavity by two capacitors C_k with matched impedance given by $Z_R = \sqrt{\frac{L}{C}} = 50\ \Omega$ the total Q factor of the resonator is modified. The small input coupling capacitors have a huge impedance compared to the loaded feedline, which is a discontinuity creating a standing wave between both C_k's. The coupling strength to the feedline can be controlled by C_k and the coupling quality factor follows as $Q_k \sim C/C_K^2 R_L \omega$. Therefore the total quality factor of the coupled cavity reads

$$\frac{1}{Q_{tot}} = \frac{1}{Q_I} + \frac{1}{Q_k} \tag{2.33}$$

and we can state a total energy dissipation rate κ as

$$\frac{1}{\kappa} = R_{tot} C_{tot} \tag{2.34}$$

with total resistance R_{tot} and capacitance C_{tot}. One should also note that the coupling capacitors also change the resonance frequency of the circuit since ω is given by $1/\sqrt{LC_{tot}}$. The ratio of coupling to internal quality factor

$$g_c = \frac{Q_I}{Q_k} = \frac{\kappa_k}{\kappa_I} \tag{2.35}$$

is an important figure of merit and one can identify three different cases for g_c:

- $g_c < 1$ the resonator is under-coupled, and Q_{tot} is dominated by Q_I and κ_I
- $g_c > 1$ the resonator is over-coupled, and Q_{tot} is dominated by Q_k and κ_k
- $g_c = 1$ the resonator is critically coupled, meaning $Q_I = Q_k$ and $\kappa_I = \kappa_k$

As already discussed the resonator is connected to a matched feedline as is shown in Fig. 2.8. The feedline further is connected to a 50 Ω load on both sides, a voltage source at the left port (number one) and to a detector on the right port (number two). Through the source an alternating voltage can be applied and is transmitted through the cavity and detected at port two. The impedance matrix Z_{ij} of the circuit relates voltages V_i and currents I_j on ports i and j by

$$Z_{ij} = \frac{V_i}{I_j} \tag{2.36}$$

with $I_j = 0$ for $j = i$. A more convenient formalism is to use scattering matrices relating input V^- and output V^- voltages at every port by the relation $V^- = SV^+$ (see Fig. 2.9). The generalized scattering parameters for the voltages at port one and two read

$$V_i = \frac{1}{Z_0}(V_i^- + V_i^+) \tag{2.37}$$

which are generally used if the characteristic impedance changes over a multiport network. The relation of voltages at both ports then follow as

$$\begin{pmatrix} V_1^- \\ V_2^- \end{pmatrix} = \begin{pmatrix} S_{11} & S_{12} \\ S_{21} & S_{22} \end{pmatrix} \begin{pmatrix} V_1^+ \\ V_2^+ \end{pmatrix} \tag{2.38}$$

where the scattering matrix describes the response of a resonator or more generally speaking a device under test (DUT).

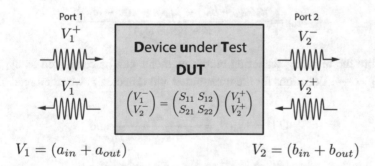

Fig. 2.9 Visualization of a device under test (DUT). A voltage at port 1 $V_1 = V_1^+ + V_1^-$ is related by a scattering matrix S to a $V_2 = V_2^+ + V_2^-$ at port 2

2.3.3   Quantum Mechanical Description of a Real LC Cavity

The resonator and its input and output signals are related by the scattering matrix of the cavity. A different treatment of this problem is known as input output formalism in which the Hamiltonian in Eq. 2.28 is used in a Master equation approach. In the Heisenberg picture an operator equation is given therefore by

$$\dot{a}(t) = -\frac{i}{\hbar}[a(t), \mathcal{H}] - (2\kappa_k + \kappa_I)a(t) + \sqrt{2\kappa_k}(a_{in}(t) + b_{in}(t)) \qquad (2.39)$$

for the cavity field operator a. The rate κ_I introduces damping by an internal reservoir. The external modes $a_{in}(t) \propto V_1^+$ and $b_{in}(t) \propto V_2^+$ are coupled to the resonator mode by $\sqrt{\kappa_k}$ according to a first Markov approximation. This linear operator equation can thus be written as

$$\dot{a}(t, \omega_p) = -i(\omega - \omega_p)a(t) - (2\kappa_k + \kappa_I)a(t) + \sqrt{2\kappa_k}(a_{in}(t) + b_{in}(t)) \quad (2.40)$$

for a symmetrically coupled two port cavity (i.e. $\kappa_k \to 2\kappa_k$) with an external driving voltage with frequency ω_p ($V = V_0(\exp(i\omega_p) + \exp(-i\omega_p))$). The frequency dependent scattering matrix is obtained by solving the cavity operator equation for the stationary state (i.e. $\dot{a}(t, \omega_p) = 0$), with the boundary conditions at each port reading

$$a_{in} + a_{out} = \sqrt{\kappa_k}\, a \text{ and } b_{in} + b_{out} = \sqrt{\kappa_k}\, a . \qquad (2.41)$$

After introducing the total dissipation rate as $\kappa = \kappa_I + \kappa_k$ one finds then

$$b_{out} = \frac{\kappa_k(a_{in} + b_{in}) - b_{in}(\kappa - i(\omega - \omega_p))}{\kappa - i(\omega - \omega_p)} . \qquad (2.42)$$

Therefore the resulting scattering matrix elements can be calculated as $S_{21} = \frac{b_{out}}{a_{in}}$ and $S_{11} = \frac{a_{out}}{a_{in}}$. Equations for the transmitted and reflected power then read

$$|T|^2 = |S_{21}|^2 = \frac{\kappa_k^2}{\kappa^2 + (\omega - \omega_p)^2} \text{ and}$$

$$|R|^2 = |S_{11}|^2 = 1 - \frac{\kappa_k^2}{\kappa^2 + (\omega - \omega_p)^2} \qquad (2.43)$$

since the average power delivered e.g. to port one is given by $P_1 = \frac{1}{2}(|a_{in}|^2 + |a_{out}|^2)$. Again three different coupling scenarios for the ratio of $g_c = \frac{\kappa_k}{\kappa_I}$ are found as introduced in the previous section and are shown in Fig. 2.10. In the discussed experiments later on over-coupled cavities are used and the dissipation rate of the resonator is approximated by a single dissipation constant $\kappa \sim 2\kappa_k$.

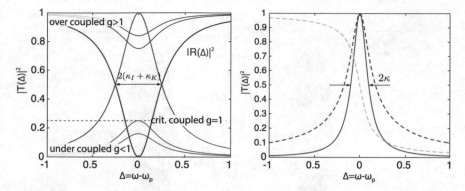

Fig. 2.10 *Left* Transmission $|T|^2 = |S_{21}|^2$ and reflection $|R|^2 = |S_{11}|^2$ of a two port *RLC* resonator. If $\kappa_I > \kappa_k$ the resonator is under coupled, critically coupled for $\kappa_I = \kappa_k$ and over coupled for $\kappa_I < \kappa_k$. *Right* The Lorentz oscillator model where only one dissipation rate κ determines the line width. *Red solid line* corresponds to the transmitted squared cavity amplitude, *blue dotted line* cavity amplitude and *yellow dotted line* to the phase of the resonator

A Hamiltonian for the coupled cavity (i.e. LC resonator) that includes a driving term then reads

$$\mathcal{H}_c = \hbar\omega_c a^\dagger a + i\hbar\eta(ae^{i\omega_p t} - a^\dagger e^{-i\omega_p t}) \tag{2.44}$$

with a cavity eigenfrequency ω_c, driving frequency ω_p and transmitted drive η. In the rotating frame one can replace ω_c with $\Delta = \omega_c - \omega_p$ and work in a frame rotating with ω_c. Dissipation can be introduced as a complex frequency contribution $-i\kappa$ to ω_c and from a Heisenberg equation

$$\dot{a} = -\frac{i}{\hbar}[a, \mathcal{H}_c] \tag{2.45}$$

the linear cavity operator equations can be derived reading $\dot{a} = (\kappa - i\Delta)a + \eta$. The steady state solution of a can be found by setting $\dot{a} = 0$ and reads

$$a = \frac{\eta}{\kappa - i\Delta}. \tag{2.46}$$

which is a complex function. The expectation value of a can be identified as the cavity field amplitude $|A|$ since $\langle a^\dagger a \rangle = |A|^2$ and an expressions for squared cavity amplitude and phase read

$$|A|^2 = \frac{\eta^2}{\Delta^2 + \kappa^2} \rightarrow |T|^2 = \frac{\kappa^2}{\Delta^2 + \kappa^2} \text{ and } \phi = \tan^{-1}\left(\frac{\Delta}{\kappa}\right) \tag{2.47}$$

with $|A|^2$ having the form of a Lorentzian function of width of 2κ (Fig. 2.11). If a coherent drive field is injected into the cavity a coherent cavity state $|\alpha\rangle$ is created

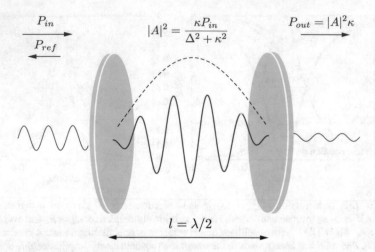

Fig. 2.11 Visualization of a the electromagnetic field between two mirrors forming a cavity. Incident P_{in} and output power P_{out} are related by the cavity and it's response can be approximate near resonance by a Lorentzian function

with a mean photon number $|A|^2 \rightarrow |\alpha|^2 = \langle N \rangle$. The mean photon number for a given input power P_{in} [W] is then obtained by replacing η^2 with $\tilde{\eta}^2 = \kappa P_{in}$ (J/s^2), from which the mean photon number in the cavity follows as $|\alpha|^2 = \tilde{\eta}^2/\hbar\omega_c\kappa^2 = P_{in}/\hbar\omega_c\kappa$ with $\Delta = 0$. Therefore the driving power needed to have on average one photon in the cavity corresponds to $\eta^2 = \kappa^2$, since $\langle N \rangle = \eta^2/\kappa^2 = 1$.

The cavity amplitude sets into a steady state under the action of a continuous drive. After the drive is switched off again the intra-cavity field $|A|^2$ decays, which is known as cavity ring down. The time dependent cavity amplitude after a coherent drive with power P_{in} is switched off is given by

$$\dot{\alpha}(t) = -\kappa|\alpha(0)| \tag{2.48}$$

with initial amplitude $\alpha(0)$ and the solution $\alpha(t) = |\alpha(0)|e^{-(\kappa+i\omega)t}$. The energy and mean photon number decays exponentially as

$$\langle N(t) \rangle = |\alpha(0)|^2 e^{-2\kappa t} \tag{2.49}$$

with a time constant $1/2\kappa$. The cavity Hamiltonian in Eq. 2.44 is solved by a Master equation approach using a quantum optics toolbox,[2] with a collapse operator $\sqrt{2\kappa}a$ and under resonant driving $\omega_p = \omega_c$. In Fig. 2.12 the time dependent mean cavity

---

[2]I will use http://qutip.org and can highly recommend this free python based Quantum optics tool box.

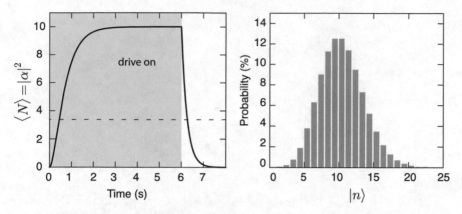

Fig. 2.12 *Left* time dependent mean cavity photon number $\langle N \rangle$ under the action of a rectangular coherent drive pulse. *Right* the corresponding photon distribution in Fock space for the cavity steady state under action of a coherent drive, derived from a full quantum mechanical calculation using http://qutip.org

photon number is plotted for $\kappa/2\pi = 2$ Hz and a drive $\eta^2 = 40$ photons/s^2. After the drive is switched on the cavity field will build up and reaches its steady state value of $\langle N \rangle = \eta^2/\kappa^2 = 10$ photons and exponentially decays with 2κ after switch off.

References

1. Jackson, *Classical Electrodynamics*, 3rd edn. (Wiley India Pvt. Limited, Jan 2007). ISBN 978-81-265-1094-8
2. S. Haroche, J.-M. Raimond, *Exploring the Quantum: Atoms, Cavities, and Photons* (OUP Oxford, Apr 2013). ISBN 978-0-19-968031-3
3. M.O. Scully, M.S. Zubairy, *Quantum Optics* (Cambridge University Press, Sept 1997). ISBN 978-0-521-43595-6
4. D. Budker, D.F. Kimball, D.P. DeMille, *Atomic Physics: An Exploration Through Problems and Solutions*. (Oxford University Press, 2004). ISBN 978-0-19-850950-9
5. R.P. Feynman, R.B. Leighton, M. Sands, *The Feynman Lectures on Physics: Mainly Mechanics, Radiation, and Heat Vol. 1* (Basic Books, 2011). ISBN 978-0-465-02493-3
6. S.M. Girvin, *Circuit QED: Superconducting Qubits Coupled to Microwave Photons*. (Oxford University Press, 2011)
7. D.F. Walls, G.J. Milburn, *Quantum Optics* (Springer Science & Business Media, Jan 2008) ISBN 978-3-540-28573-1
8. D.M. Pozar, *Microwave Engineering*. (Wiley, Feb 2004). ISBN 978-0-471-44878-5
9. C. Cohen-Tannoudji, B. Diu, F. Laloe, *Quantum Mechanics, Vol. 1 & 2*, 1st edn. (Wiley, New York, June 1977). ISBN 978-0-471-16433-3

Chapter 3
Spins in the Cavity—Cavity QED

In this chapter I will focus on the brief description of the interaction between a single cavity field mode and single or ensembles of electron spins. Starting from the simplest case of a single two level system in the cavity, I introduce a formalism to treat ensembles of spins coupling to the cavity field mode. Explaining from single spin interactions the principle of collective enhancement in light and matter interaction. For an in depth treatment of the following phenomena and the broad field of cavity QED I would like to emphasize the textbooks [1–3] on which the following sections are partly based on.

3.1 Single Spin in the Cavity

After having derived a Hamiltonian for the cavity field mode in the previous chapter a single spin is placed inside the cavity mode volume. The Hamiltonian describing the system of a spin inside a cavity including an interaction of both systems reads then

$$\mathcal{H} = \mathcal{H}_{cav} + \mathcal{H}_{spin} + \mathcal{H}_{int} \tag{3.1}$$

with $\mathcal{H}_{cav} = \hbar\omega_c a^\dagger a$ and $\mathcal{H}_{spin} = \frac{\hbar\omega_s}{2}(|\downarrow\rangle\langle\downarrow| - |\uparrow\rangle\langle\uparrow|) = \frac{\hbar\omega_s}{2}\sigma_z$. The last term describes the interaction of the electromagnetic field inside the cavity with the magnetic or electric moment of an electron spin. As indicated by the Pauli matrices σ_z a single electron spin is treated as a quantum-mechanical two-level system with two eigenstates $|\downarrow\rangle$ and $|\uparrow\rangle$. The two-level system can be considered as the purest possible anharmonic system while the cavity is a truly harmonic system.

In this thesis, the main focus lies on the interaction of the magnetic field of the cavity with the spin magnetic dipole moment (Fig. 3.1) of an electron S which can be written as

© Springer International Publishing AG 2017

S. Putz, *Circuit Cavity QED with Macroscopic Solid-State Spin Ensembles*,
Springer Theses, DOI 10.1007/978-3-319-66447-7_3

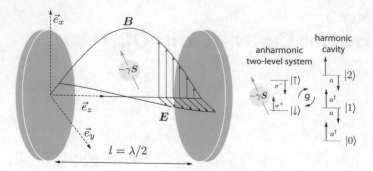

Fig. 3.1 Visualization of a single spin S with a magnetic moment μ inside a single mode cavity with electric E and magnetic B field components. The magnetic field of the harmonic oscillator couples to the magnetic dipole moment of the anharmonic spin two-level system. The interaction is denoted by the term $-\mu S$ in the total Hamiltonian describing the coupled cavity spin system

$$\mu = -\frac{g_e \mu_B S}{\hbar} = -\gamma_e S \tag{3.2}$$

with the electron g_e factor, the Bohr magneton μ_B and gyromagnetic ratio γ_e. The interaction term of the Hamiltonian in Eq. 3.1 therefore reads

$$\mathcal{H}_{int} = -\mu B \tag{3.3}$$

with the magnetic field B produced by the mode inside the cavity. This is the most basic principle on which all physical phenomenas of this thesis are based on. In the following sections the interaction will be discussed, going from a single spin to an ensemble of N non interacting spins in the cavity. This should give the reader the ability to understand the underlying principles of the experiments carried out in the experimental chapters of this thesis.

3.1.1 Cavity Spin Interaction

The principle mechanism of the cavity spin interaction was already introduced and in the latter the interaction will be treated in a full quantum mechanical manner, which is known as cavity QED in physics. Hence the magnetic dipole moment is treated as a single two-level system with two eigenstates $|i\rangle = |\uparrow\rangle$ and $|j\rangle = |\downarrow\rangle$. The magnetic moment of such a spin two-level system ($|\downarrow\rangle, |\uparrow\rangle$) can then be written as

$$\mu = -\gamma_e \sum_{ij} |i\rangle\langle i|\sigma|j\rangle\langle j| \tag{3.4}$$

using spectral decomposition. With the dipole transition matrix element $M_{ij} = \langle i|\boldsymbol{\sigma}|j\rangle$ and $\sigma_{ij} = |j\rangle\langle i|$ one finds

$$\boldsymbol{\mu} = -\gamma_e \sum_{ij} M_{ij}\sigma_{ij} .\qquad(3.5)$$

The interaction Hamiltonian therefore reads

$$\mathcal{H}_{int} = \mathrm{i}\gamma_e B_0 \sum_{ij} M_{ij}\sigma_{ij}(a - a^\dagger)\qquad(3.6)$$

with the quantized magnetic field mode $\boldsymbol{B} = -\mathrm{i}B_0(a - a^\dagger)$ (see Sect. 2.2.3). If we evaluate this Hamiltonian we find the interaction term

$$\mathcal{H}_{int} = \mathrm{i}\hbar g(\sigma_+ + \sigma_-)(a - a^\dagger)\qquad(3.7)$$

with a single spin interaction strength $g_{ij} = \frac{\gamma B_0 M_{ij}}{\hbar} \equiv g, \sigma_+ = \sigma_{ij}$ and $\sigma_- = \sigma_{ji}$. The spin raising and lowering operators $\sigma_\pm = \sigma_x \pm \mathrm{i}\sigma_y$ allow then to create $|\uparrow\rangle = \sigma_+|\downarrow\rangle$ or $|\downarrow\rangle = \sigma_-|\uparrow\rangle$ destroy a spin excitation.

The product of Eq. 3.7 includes terms $\sigma_- a$ and $\sigma_+ a^\dagger$ which are processes in which the energy is not conserved, since in the first case the spin makes a transition from the excited to the ground state, while a photon is annihilated in the cavity (Fig. 3.2). In the second case a spin goes from the ground in the excited state while a photon is created in the cavity. Such processes result in the gain or loss of two photons (i.e. two photon processes) and the energy of $2\hbar\omega$, respectively. These terms are usually dropped, known as the so called rotating-wave approximation (RWA), by ignoring counter rotating terms which can occur in Eq. 3.7. This approximation is only valid if the interaction strength g is small compared to the transition frequency ω.

Fig. 3.2 Eigenenergies observed from diagonalizing the Jaynes-Cummings Hamiltonian in Eq. 3.8 before making the rotating wave approximation (RWA). By removing terms $a\sigma_-$ and $a^\dagger\sigma$ from the interaction part of the Hamiltonian the model is restricted to only one photon processes which corresponds to the rotating wave approximation

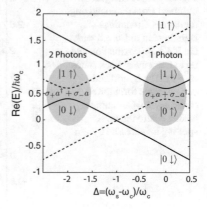

3.1.2   The Jaynes Cummings Model

The total Hamiltonian, after introducing the interaction term describes the coupled cavity spin system, reads in the RWA approximation

$$\mathcal{H} = \hbar\omega_c a^\dagger a + \frac{1}{2}\hbar\omega_s \sigma_z + i\hbar g(\sigma_+ a - \sigma_- a^\dagger) \tag{3.8}$$

to which a cavity driving term $i\hbar\eta(a - a^\dagger)$ will be added in the latter. This Hamiltonian can be diagonalized which gives the eigenstates and energies for the coupled system. In Fig. 3.3 the solution of the diagonalized Hamiltonian including up to three cavity Fock states is plotted. Avoided crossing are observed between coupled states $|Cavity, Spin\rangle$ in which cavity and spins exchange one quanta of energy, for example between states $|1 \downarrow\rangle$ and $|0 \uparrow\rangle$. If the cavity transition frequency ω_c is equal to the spin transition frequency ω_s this states hybridize and a new class of states $|n\pm\rangle$ known as polariton modes or dressed states are formed. The splitting Ω between the coupled cavity spin system is often refereed to as a normal-mode or Rabi splitting and scales as $\Omega = g\sqrt{n}$ with n the number of excitations in the cavity. The hybridized energy levels in the resonant case then form the so called Jaynes-Cummings ladder, where the level splitting increases if the ladder is climbed. The Rabi frequency changes when the number of photons in the cavity is increased, but compared to the semi classical Rabi model the Jaynes-Cummings model is a full quantum mechanical treatment of the spin and electromagnetic field.

As discussed in the ideal case of a single two-level spin coupled to a linear resonator the system is described in the framework of the Jaynes-Cummings model. This is a convenient treatment but in the "real" world spins are often multi-level systems, which are only considered to be effective two-level systems. It is therefore

Fig. 3.3 Eigenenergies observed from diagonalizing the Jaynes-Cummings Hamiltonian in Eq. 3.8 with the RWA approximation and up to four cavity Fock states $|0 - 3\rangle$. The hybridized energy levels form the so called Jaynes-Cummings ladder, where the level splitting is increased by $\sqrt{n}$

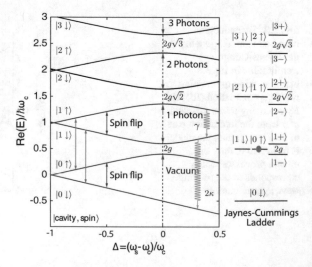

instructive to have a brief look at the solution of an Hamiltonian similar to Eq. 3.8 but solved for a three-levelsystem $S = 1$, $m_s = \downarrow, 0, \uparrow$. Furthermore it is assumed that the spin triplet is described by a Hamiltonian $\mathcal{H}_{spin} = hDS_z^2 + h\mu BS_z$ which features a zerofield D and Zeeman B splitting term. The considered spin Hamiltonian with a rather large zerofield splitting term ($2\pi D \approx \omega_c$) will also apply to the spin system used later on in this thesis. In Fig. 3.4 the detailed solution is shown following from diagonalizing the cavity spin Hamiltonian for up to three cavity Fock states. The level structure is considerably more complicated. We see multiple eigenstates $|Cavity, Spin\rangle$ hybridizing as coupled cavity spin states

$$|\pm\rangle_1 \sim \frac{1}{\sqrt{2}}(|10\rangle \pm |0\uparrow\rangle)$$

$$|\pm\rangle_2 \sim \frac{1}{\sqrt{2}}(|1\downarrow\rangle \pm |00\rangle) \qquad (3.9)$$

depending on the cavity frequency ω_c and the frequency of both spin transitions $\omega_\uparrow$ and $\omega_\downarrow$. Naturally the occurrence of $|\pm\rangle_1$ and $|\pm\rangle_2$ depends on which transition $\omega_\uparrow$ and $\omega_\downarrow$ is resonant with the cavity. In order to treat this three-level system as an effective two-level system both spin transition have to be well separated with respect to the cavity frequency and interaction strength. If the spectral distance of both spin transitions becomes too small this gives rise to spin level mixing and the influence of nearby spin energy levels can not be neglected. This also means a well defined level separation has to be ensured in the experiments in order to justify the validity of the assumption of working with an effective two-level system.

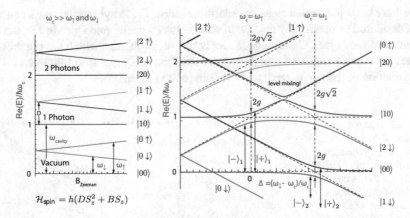

Fig. 3.4 Eigenenergies observed from diagonalizing the Jaynes-Cummings Hamiltonian for a single three-level system with the RWA approximation and three cavity Fock states $|0-2\rangle$. Four possible polariton modes $|\pm\rangle_1$ and $|\pm\rangle_2$ are observed and only if the transition separation is large enough one can treat the spintriplet as effective two-level systems

3.1.3   Dressed States/Polariton Modes

As was shown, the complete level structure of the full Jaynes-Cummings Hamiltonian is rather complicated. Therefore from a practical point of view, only coupled cavity and spin resonance transitions will be treated by reducing the eigenbasis of the Hamiltonian in Eq. 3.8 to

$$\mathcal{H} = \hbar \begin{pmatrix} \omega_c & -ig \\ ig & \omega_s \end{pmatrix} \tag{3.10}$$

with the restriction of maximally one photon in the cavity. By solving the eigenvalue problem $\mathcal{H}|\pm\rangle - \lambda|\pm\rangle = 0$ the eigenenergies of the coupled system follow as

$$\mathcal{E}_\pm = \frac{\hbar}{2}(\omega_s + \omega_c \pm \sqrt{4g^2 + \Delta^2}) \tag{3.11}$$

with a cavity spin frequency detuning $\Delta = \omega_s - \omega_c$. The eigenstates of the coupled system $|\pm\rangle$ are a superposition of the cavity and spin eigenstates $|1 \downarrow\rangle$ and $|0 \uparrow\rangle$, respectively. This newly formed states can be paramterized as $|+\rangle = \cos(\theta)|1 \downarrow\rangle + \sin(\theta)|0 \uparrow\rangle$ and $|-\rangle = \sin\theta|1 \downarrow\rangle - \cos\theta|0 \uparrow\rangle$ with an angle θ given by $\tan^{-1}(g/\Delta)$. If the detuning Δ is zero these eigenstates share equal contributions from both spin and cavity states carrying a single excitation

$$|\pm\rangle = \frac{1}{\sqrt{2}}(|1 \downarrow\rangle \pm |0 \uparrow\rangle) \tag{3.12}$$

which are know as dressed states or polariton modes. These hybridized states can also be considered as maximally entangled state between cavity and spin. In Fig. 3.5 the solution of the Jaynes-Cummings model is shown. Both lines which are the dressed states $|\pm\rangle$ show an avoided crossing with a splitting of $2g$ and the color encodes the contribution of the cavity *(red)* and the spin *(black)* eigenstates.

Fig. 3.5 Dressed state picture of the coupled cavity spin system. Whether an eigenstate is spin *black* or cavity *red* like is indicated by the color gradient. On resonance $\omega_s = \omega_c$ the system hybridizes to the dressed states $|\pm\rangle$ which are an equal superposition of cavity and spin and are called polariton modes

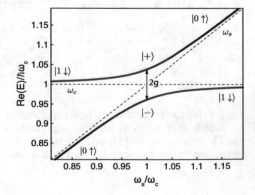

As the cavity and spin detuning Δ is increased both states $|\pm\rangle$ will become almost decoupled cavity and spin eigenstates. However, for very large detunings one can see that the states $|\pm\rangle$ are lifted from their bare and uncoupled transition frequency ω_s and ω_c. This is known as a dispersive shift in cavity QED and the energy level shift can be estimated by making a Taylor expansion of Eq. 3.11

$$\mathcal{E}_{\text{disp}} \approx \hbar(\omega_c \pm \frac{g^2}{\Delta}) \tag{3.13}$$

from which the shifted energy of the bare cavity resonance follows. This approximation is valid if the detuning $\Delta \gg g$ which is known as the dispersive regime.

In a next step dissipation channels for the cavity and spin system are added to the ideal Jaynes-Cummings model. As shown in the previous section this can be done by introducing complex frequency parts, being the cavity κ and spin γ dissipation rates.[1]

The modified Hamiltonian of Eq. 3.10 then reads

$$\mathcal{H} = \hbar \begin{pmatrix} \omega_c + i\kappa & -ig \\ ig & \omega_s + i\gamma \end{pmatrix} \tag{3.14}$$

and one finds an imaginary contribution in the eigenenergies of the hybridized states $|\pm\rangle$

$$\mathcal{E}_{\pm} = \frac{\hbar}{2}(\omega_s + \omega_c - i(\kappa + \gamma) \pm \sqrt{4g^2 - (\gamma - \kappa)^2}) \tag{3.15}$$

for the resonant case $\Delta = 0$. The total dissipation rate of the system then follows as the imaginary part of the eigenenergy and reads

$$\Gamma = \frac{\kappa + \gamma \pm \text{Re}\{\sqrt{(\gamma - \kappa)^2 - 4g^2}\}}{2} \tag{3.16}$$

with the eigenfrequencies of the dressed states

$$\Omega_R = \text{Re}\{\sqrt{(4g^2 - (\gamma - \kappa)^2)}\} \tag{3.17}$$

giving rise to the Rabi splitting with the Rabi frequency Ω_R. For the resonant case ($\Delta = 0$) and if $\Omega_R \geq g$ the total system decoherence rate then reaches the natural limit of

$$\Gamma = \frac{\kappa + \gamma}{2} . \tag{3.18}$$

---

[1]Nota bene: for the sake of simplicity all dissipation rates in this chapter are introduced as the half-width at half-maximum of the corresponding natural linewidth. In the experimental chapters the measure for each rate observed will be indicated.

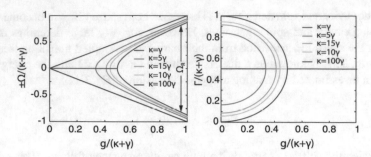

Fig. 3.6 *Left* Rabi splitting Ω_R (see Eq. 3.17) for different relations of κ and γ. *Right* Total dissipation rate Γ (see Eq. 3.16) for the same relations of κ and γ used to calculate Ω_R. As a critical value of $\frac{g}{\kappa+\gamma} = 0.5$ is reached, where the mean dissipation rate is equal to the coupling strength $\Gamma = \frac{\kappa+\gamma}{2}$, the system enters the strong coupling regime

In this limit a polariton mode $|\pm\rangle$ shares an excitation equally between cavity and spin eigenstate and therefore to total system decay rate Γ is given and bound from below by the mean dissipation rates of both individual subsystems. From Eq. 3.17 the relation of κ, γ and g allows to identify a crucial limit, which is known as the strong coupling regime and is reached if κ and $\gamma \leq g$, as is shown in Fig. 3.6.

As dissipation is introduced the system polariton modes exhibit a finite linewidth Γ. If an additional driving term with frequency ω_p is included, the cavity transmission response or scattering matrices can be calculated from

$$T = \text{Tr}\{\text{inv}(\omega_p \mathbf{1} - \mathcal{H})\}\,|_{ij=11} \ . \tag{3.19}$$

For a single spin in the cavity mode volume a complex function then follows for the system transmission

$$T = i\kappa \frac{\omega_p - \omega_s - i\gamma}{-g^2 + (\omega_p - \omega_s - i\gamma)(\omega_c - \omega_p + i\kappa)} \tag{3.20}$$

with the coupling strength g, the spin and cavity transition frequencies ω_s and ω_c, respectively. In Fig. 3.7 the transmission $|T|^2 = |S_{21}|^2$ is plotted as a function of δ_c, for spin and cavity on resonance ($\Delta = \omega_s - \omega_c = 0$) and different coupling strengths g. For a coupling strength $g = 0$ the bare cavity transmission is observed. If the spin cavity interaction increases this gives rise to the so called vacuum Rabi splitting. We can distinguish two fundamentally important regimes:

- **The weak coupling regime:** for $g \ll \frac{\kappa+\gamma}{2}$ the cavity spin interaction is not strong enough to overcome the dissipation rates κ and γ, but the bare cavity response eventually is smeared out. If the cavity dissipation rate is the dominating decay rate $\kappa \gg \frac{g^2}{\gamma} \gg \gamma$ this is known as the bad cavity regime.
- **The strong coupling regime:** for $g \geq \frac{\kappa+\gamma}{2}$ a vacuum Rabi splitting or normal mode splitting is observed. If the coupling strength reaches a critical value of

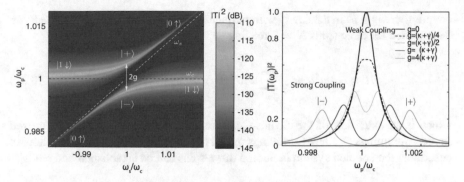

Fig. 3.7 *Left* Dressed state picture of the coupled cavity spin system. On resonance $\omega_s = \omega_c$ the system hybridizes to the dressed states or polariton modes $|\pm\rangle$. *Right* Transmission spectra on resonance. If cavity κ and spin dissipation rate γ are small compared to the coupling strength g the system eventually enters the strong coupling regime

$\frac{g}{\kappa+\gamma} = 0.5$ the system makes a transition from the weak to the strong coupling regime. A figure of merit is the system cooperativity which fulfills in the strong coupling regime $C = \frac{g^2}{\kappa\gamma} \geq \frac{(\gamma+\kappa)^2}{4\gamma\kappa} = \frac{1}{2} + \frac{1}{4}(\frac{\gamma}{\kappa} + \frac{\kappa}{\gamma})$. One should also note that a cooperativity of $C \geq 1$ is only sufficient to reach the strong coupling regime when $\kappa \approx \gamma$.

3.2 Ensembles of Spins in the Cavity

The physics of a single spin coupled to the cavity can be treated nicely in the Jaynes-Cummings framework, as we have seen in the previous sections. For a single two level system the coupling to the cavity magnetic field gives rise to two spin dressed states $|\pm\rangle$ and eigenstates. In the case of more than one two-level system coupled to the cavity, additional eigenstates are created which need to be considered. In the following two new classes of eigenstates of very different and distinct nature will be introduced. These states are called "bright" and "dark" states or "superradiant" and "subradiant", respectively, corresponding to their interaction strength with the cavity mode. The formation and their physical properties will be derived and discussed from an intuitive model in the latter.

3.2.1 Three Spins in the Cavity

In a simplified model the creation of dark and subradiant states can be discussed by treating the case of having three spins $j = \{1, 2, 3\}$ with transition frequency ω_j and coupling strength g_j in the cavity with frequency ω_c. The Hamiltonian in Eq. 3.8 can be extended for three spins $N = 3$ reading

$$\mathcal{H} = \hbar\omega_c a^\dagger a + \frac{\hbar}{2}\sum_{j=1}^{N}\omega_j\sigma_j^z + \hbar\sum_{j=1}^{N}g_j\left[\sigma_j^- a^\dagger + \sigma_j^+ a\right] \qquad (3.21)$$

which is an extension of the Jaynes-Cummings model known as the generalized Jaynes-Cummings or Tavis-Cummings model [4]. Similar to Eq. 3.14 a Hamiltonian describing the coupled system including three spins can be established and reads

$$\mathcal{H} = \hbar\begin{pmatrix} \omega_c + i\kappa & -ig_1 & -ig_2 & -ig_3 \\ ig_1 & \omega_1 + i\gamma_1 & 0 & 0 \\ ig_2 & 0 & \omega_2 + i\gamma_2 & 0 \\ ig_3 & 0 & 0 & \omega_3 + i\gamma_3 \end{pmatrix} \qquad (3.22)$$

with a single cavity Fock state. In the case of having an individual spin detuning much larger than the individual coupling strengths $\Delta_{i,k} = \omega_i - \omega_k \gg g_i$ the dressed states, similar to to the Jaynes-Cummings model, read

$$|\pm\rangle_1 \sim \frac{1}{\sqrt{2}}(|1\downarrow\downarrow\downarrow\rangle \pm |0\uparrow\downarrow\downarrow\rangle)$$

$$|\pm\rangle_2 \sim \frac{1}{\sqrt{2}}(|1\downarrow\downarrow\downarrow\rangle \pm |0\uparrow\downarrow\downarrow\rangle) \qquad (3.23)$$

$$|\pm\rangle_3 \sim \frac{1}{\sqrt{2}}(|1\downarrow\downarrow\downarrow\rangle \pm |0\uparrow\downarrow\downarrow\rangle)$$

which is of course a coarse simplification and ignores any dispersive shift by proximate spin levels. The Hamiltonian in Eq. 3.22 can be diagonalized by solving the eigenvalue problem $\mathcal{H}|\Psi\rangle - \lambda|\Psi\rangle = 0$ and the real part $\text{Re}(\lambda)$ corresponding to the eigenenergies of the coupled system is plotted in Fig. 3.8 for different spin detunings $\omega_i - \omega_j$.

The new eigenstates can be expressed in the basis of the uncoupled states and the spin cavity detunings determine the contribution to each individual dressed state. The eigenstates for non-degenerated and largely separated spins are shown in the upper left panel of Fig. 3.8. For large detunings ($\Delta_{1,2,3} = \omega_{1,2,3} - \omega_c \gg g$) cavity and spins remain in their uncoupled eigenstates but face a dispersive shift of the order of $\approx g_j^2/\Delta_{1,2,3}$. In the frequency region exactly between spin one and two this shift is canceled and the corresponding eigenstate reads

$$|X\rangle = -A|0\uparrow\downarrow\downarrow\rangle + B|1\downarrow\downarrow\downarrow\rangle + C|0\downarrow\uparrow\downarrow\rangle \qquad (3.24)$$

with amplitudes A, B and C. Hence in the scenario of large spin detunings one can assume $B \approx 1$ and $A \approx 0$, $C \approx 0$. An interesting phenomena happens if the spin detunings $\Delta_1 = \omega_1 - \omega_2$ and $\Delta_2 = \omega_2 - \omega_3$ become small and the cavity is in

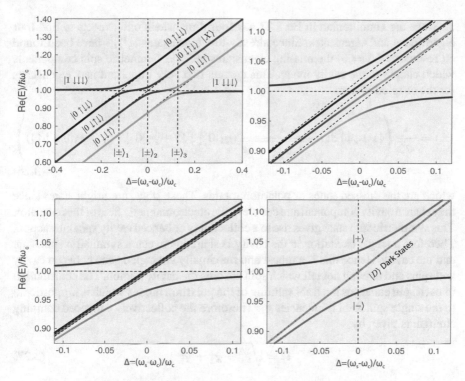

Fig. 3.8 *Upper left* Three largely detuned spins in the cavity mode volume. *Upper right and lower left* As the spins are brought close to resonance with each other the eigenstates start to mix. *Lower right* If the spins are degenerate all three spins hybridize to two collective enhanced polariton modes with a symmetrical spin state, where as the reaming state will from $N-1$ dark states being decoupled entirely from the cavity mode

resonace with the central spin. The cavity contribution to the state $|X\rangle$, proportional to B, would vanish and simultaneously the spin components, A and C, are increased. In the case of having all three spins in resonance (i.e. $\Delta_1 = \Delta_2 = 0$) state $|X\rangle$ looses completely its cavity part and one finds two new eigenstates

$$|D\rangle_1 = \frac{1}{\sqrt{1 + (\frac{g_2}{g_1})^2}}(|0 \downarrow\uparrow\downarrow\rangle - \frac{g_2}{g_1}|0 \uparrow\downarrow\downarrow\rangle)$$

$$|D\rangle_2 = \frac{1}{\sqrt{1 + (\frac{g_3}{g_2})^2}}(|0 \downarrow\downarrow\uparrow\rangle - \frac{g_3}{g_2}|0 \downarrow\uparrow\downarrow\rangle) \tag{3.25}$$

which we gave the capital letter D for dark state. Dark in this context means it is a state which couples not to the electromagnetic field of the cavity. These are partially anti symmetric under exchange of the two spins sharing one excitation.

Since the Hamiltonian in Eq. 3.22 has four dimensions one expects to find four eigenvalues and eigenstates. Since already tow eigenstates $|D\rangle_{1,2}$ have been found on resonance ($\Delta = 0$) the missing eigenstates include symmetric spin components which couple to the cavity mode. This creates two very important new eigenstates for the coupled system

$$|\pm\rangle = \frac{1}{\sqrt{2}} \left(|1 \downarrow\downarrow\downarrow\rangle \pm \frac{1}{\sqrt{g_1^2 + g_2^2 + g_3^2}} (g_1|0 \uparrow\downarrow\downarrow\rangle + g_2|0 \downarrow\uparrow\downarrow\rangle + g_3|0 \downarrow\downarrow\uparrow\rangle)) \right)$$

(3.26)

which are the dressed states or polariton modes. These states are bright states since they share a cavity component and couple to the electromagnetic field of the resonator. This symmetric spin state gives rise to a collectively enhanced cavity spin interaction. Therefore a single excitation in the cavity is shared between a symmetric spin state and the cavity. Hence this is a robust and maximally entangled state between cavity and spins and one can not tell which spin absorbs the cavity photon. The contribution of each spin enhances the Rabi splitting of the polariton modes which is proportional to the single spin Rabi frequencies g_j. Therefore the collectively enhanced coupling strength is given by

$$\Omega = \sqrt{g_1^2 + g_2^2 + g_3^2}$$

(3.27)

and one finds the real part of the eigenvalues of Eq. 3.22 for $\omega_s = \omega_{1,2,3}$ as

$$\text{Re}\{\omega_{\text{Dark}}\} = \Delta \quad \text{and} \quad \text{Re}\{\omega_{\text{Bright}}\} = \frac{1}{2}(\Delta \pm \sqrt{4(g_1^2 + g_2^2 + g_3^2) + \Delta^2}) .$$

(3.28)

if $\Omega \gg \kappa$ and $\gamma_{1,2,3}$. For the case of $\Delta = 0$ then follows $\omega_{\text{Bright}} = \pm\Omega = \pm\sqrt{g_1^2 + g_2^2 + g_3^2}$ as expected from looking at the new eigenstates and how the spins share a single common excitation.

A very important fact can be understood if one focuses on the imaginary part of ω_{Dark} and ω_{Bright}. By diagonalizing the Hamiltonian the imaginary parts of the eiganvalues for $\Delta = 0$ account for the dissipation in the system and read

$$\text{Im}\{\omega_{\text{Dark}}\} = \gamma \quad \text{and} \quad \text{Im}\{\omega_{\text{Bright}}\} = \frac{1}{2}(\gamma + \kappa)$$

(3.29)

if $\Omega \gg \kappa$ and $\gamma = \gamma_{1,2,3}$. As one can see the total decay rate of the strongly coupled system and collective enhanced bright state $|B\rangle$ is still described by the formula derived earlier for the Jaynes-Cummings model as

$$\Gamma = \frac{\kappa + \gamma}{2} .$$

(3.30)

Due to this fact the collectively enhanced coupling scales up with the square root of spins in the cavity while the total dissipation rate for the system remains constant. This also means that if the single spin interaction strength is not strong enough to ensure strong coupling, collective enhancement allows to enter the strong coupling regime although $g \ll \kappa$ and γ.

After treating the situation of a rather small ensemble of three spins in the cavity sharing a single excitation one can extend the number of cavity Fock states up to where the number of photons may become equal to the number of spins. The maximal number of photons which can be absorbed by three spins is of course three. This means in order to describe the full eigenenergy spectrum of an inverted ensemble one has to include up to four cavity Fock states. It is then convenient to reduce these three spins to a single giant spin with spin quantum number $S = 3/2$ and construct the full sub space of eigensates as shown in Fig. 3.9. One should note that the derived dark spin states $|D\rangle_{1,2}$ are not eigenstates of $S = 1/2$ spin states but rather a superposition of the actual dark states which must fulfill $(S = 1/2)$ as is shown Fig. 3.9. Although $|D\rangle_{1,2}$ would be "dark" with respect to the cavity this is a rather subtle fact but can be understood since $|D\rangle_{1,2}$ miss a third spin component and $\langle D_1|D_2\rangle \neq 0$. Also more then the stated two spin combinations can be found that fulfill $\langle B|D_{1,2}\rangle = 0$. In contrast only two dark states that are eigenstates of $S = 1/2$ and fulfill $\langle B|D_{1,2}\rangle = 0$ and $\langle D_1|D_2\rangle = 0$ can be found which fully defines the spin states of the coupled system. This will become an important aspect in the following section and correct expression for the spin dark states will be introduced, but the intuitively derived formalism for the spin bright states still holds true.

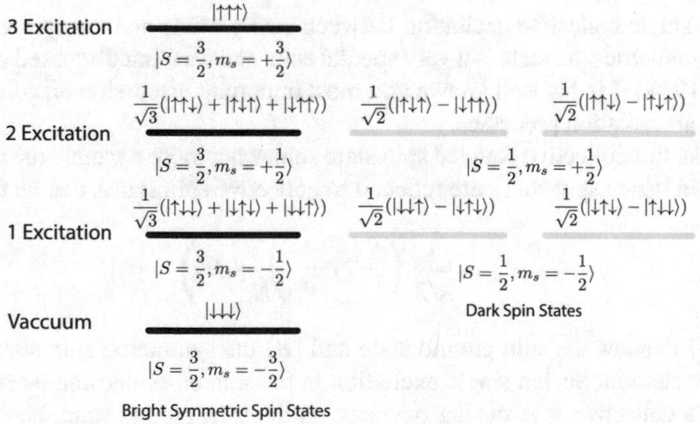

Fig. 3.9 If the cavity Fock space is increased from $|0\rangle$ and $|1\rangle$ to $|3\rangle$ the three resonant spin states can be treated as one giant spin with $S = 3/2$ and the derived eigenenergy spectrum for symmetrical superradiant (*bright*) and subradiant (*dark*) states are shown

3.2.2 N Spins in the Cavity

The Tavis-Cummings model was discussed in the simple but very instructive regime of having only three spins in the cavity. From which bright and dark states were derived and the collectively enhancement of the cavity spin interaction was introduced. This formalism will now be extended to an arbitrary number of N spins interacting with the electromagnetic field of the cavity mode. If one looks at Eq. 3.27 the collectively enhanced coupling strength thus can be extended and reads

$$\Omega = \sqrt{\sum_j^N |g_j|^2} \,. \tag{3.31}$$

From the assumption of a uniform single spin coupling strengths (Rabi frequencies) $g_j = g$ follows that the coupling strength

$$\Omega = \sqrt{N} g \tag{3.32}$$

scales as with the number of spins N as $\sqrt{N}$. The bright states or polariton modes read under the same consideration then

$$|\pm\rangle = \frac{1}{\sqrt{2}} \left(|1 \downarrow\downarrow \ldots \downarrow\rangle \pm \frac{1}{\sqrt{N}} |0\rangle \left(|\uparrow\downarrow \ldots \downarrow\rangle_1 + |\downarrow\uparrow \ldots \downarrow\rangle_2 + \cdots + |\downarrow\downarrow \ldots \uparrow\rangle_N \right) \right) \tag{3.33}$$

sharing a single collective excitation between cavity mode and the spin ensemble. This N symmetric spin state is a very special state and was first discussed by Robert Dicke in 1954 [5] in his well known and most important discussion on coherence in spontaneous radiation processes.

To make this collective coupled spin state somewhat more tractable the following notation, in which spin states are reduced to collective spin states, can be used

$$|\pm\rangle = \frac{1}{\sqrt{2}} \left(|1G\rangle \pm \frac{1}{\sqrt{N}} |0B\rangle \right) \tag{3.34}$$

where $|G\rangle$ denotes the spin ground state and $|B\rangle$ the symmetric spin state sharing a single excitation. Such a single excitation in the spin ensemble can be created by applying a collective spin raising operator on the spin ground state, hence $|B\rangle = J_+|G\rangle$ which will be discussed in more detail in the following. With this collective spin operators

$$J_z = \frac{1}{2} \sum_N^j \sigma_z^j \text{ and } J_\pm = \sum_j^N \frac{g_j}{\sqrt{\sum_j^N |g_j|^2}} \sigma_\pm^j \tag{3.35}$$

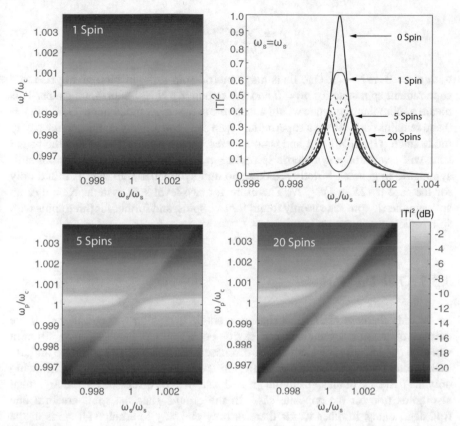

Fig. 3.10 If a single spin coupling strength g might not be sufficient to overcome cavity and spin dissipation rates κ and γ, one can use an ensemble of N of spins to collectively enhance the spin cavity interaction strength which scales as $\Omega = \sqrt{N}g$ and $\Omega \gg \kappa, \gamma \gg g$

one can rewrite the Tavis-Cummings Hamiltonian in terms of collective spin operators as

$$\mathcal{H} = \hbar\omega_c a^\dagger a + \hbar\omega_s J_z + i\Omega (J_+ a - J_- a^\dagger) \tag{3.36}$$

assuming $\Omega = \sqrt{N}g$ with $J_\pm = \frac{1}{\sqrt{N}} \sum_j^N \sigma_\pm^j$. The collective enhancement of Ω can be seen in Fig. 3.10 where the cavity transmission is shown. One clearly sees, that by increasing the number of spins the limitation of $g \ll \kappa$ and γ can be overcome and the system advances from the bad cavity regime into the strong coupling regime.

In such an ensemble, sharing a single excitation, one finds one symmetric spin state $|B\rangle$, giving rise to two polariton modes $|\pm\rangle$, and $N-1$ dark states $|D\rangle$. Super (bright) and subradiant (dark) spin states can be created by applying J_+ onto the ensemble ground state $|G\rangle = |\downarrow \ldots \downarrow\rangle$ as

$$|\Psi_m\rangle = \sum_{j=0}^{N-1} e^{im2\pi j/N} J_+|G\rangle \tag{3.37}$$

with $m = 0 \ldots N - 1$. One finds a symmetric spin superposition giving rise to a superradiant spin state for $m = 0$ and the remaining $N - 1$ subradiant states. This means a subradiant spin state would not couple to the cavity mode since $\langle G|J_-|\Psi_1\rangle = 0$ and remains dark. While a superradiant spin state collectively couples to the cavity mode since $\langle G|J_-|\Psi_0\rangle = 1$ and is considered bright. As a matter of fact the bright state will always be a symmetric spin state, but a subradiant spin state is not anti-symmetric in general. Strictly speaking an anti-symmetric spin state is found only for the case if $(2S + 1) \geq N$. Therefore for two-level spin systems $S = 1/2$ an anti-symmetric spin state is only found for two spins, and further for three spins with spin triplets $S = 1$ and so fourth.

3.2.3   The Dicke Model

So far only the case of N spins sharing a single excitation was considered. If the number of possible excitations and cavity Fock states is increased two different situations which are equivalent can be discussed; the first when all spins are initially in the ground state and the second if all spins are in the excited state. In the following only the first case of all spins in the ground state will be discussed but the argument also holds true for the opposite case. In the ground state the spins are in a one fold degeneracy, in other words there is only one way to arrange all spins in the ground state. If the ensemble is then excited with a single excitation one finds a N fold degeneracy including one symmetrical superradiant and $N - 1$ subradiant spin states. If the number of excitations is subsequently raised the number of possible permutations of arranging spin states $\uparrow$ and $\downarrow$ follow as

$$\frac{N!}{e!(N-e)!} \tag{3.38}$$

for a given number of excitation $e = \{0, .., N\}$. This formalism follows the Dicke model of treating such an ensemble of N spins as a giant spin with spin quantum number $J = N/2$ and $m_J = \pm N/2$. One should note that in the original Dicke model no cavity mode is occurring and spin correlations arise since the distance between two spins is assumed to be smaller than the wave length given by their transition frequencies.

The ground state of this giant spin corresponds then to $|N/2, -N/2\rangle = |\downarrow \ldots \downarrow\rangle = |G\rangle$ and if a collective spin operators is applied

$$J_\pm|J, m_J\rangle = \hbar\sqrt{J(J+1) - m_J(m_J \pm 1)}|J, m_J \pm 1\rangle \tag{3.39}$$

the corresponding eigenvalues can be calculated. The transition matrix elements for going from the ground to the first excited state follow as

$$M_{G \to B_1} = |B_1|J_+|G\langle = \sqrt{N} \tag{3.40}$$

which corresponds the the collectively enhanced coupling strength $\Omega = \sqrt{N}g_0$ as introduced earlier. The degeneracy of the first excited level $|G + 1\rangle$ follows from Eq. 3.38 and one finds an N fold degeneracy in the one excitation manifold of such a giant spin J. This N fold degeneracy consists of one symmetrical spin state and $N - 1$ subradiant spin states. Subradiant states can be found by assuming $J' = N/2 - 1$ with $m_J = N/2 - 1$. Since states $J = N/2$ and $J' = N/2 - 1$ are orthogonal states the set of subradiant spin states $|D\rangle \to |J = N/2 - 1, m_J\rangle$ and the symmetrical spin state $|B\rangle \to |J = N/2, m_J\rangle$ are decoupled from each other hence $\langle B|D\rangle = 0$.

If the number of excitations e is increased further all symmetrical spin states can be created by repetitively applying the collective spin operator J_+ to the spin ground state $|G\rangle = |\downarrow \ldots \downarrow\rangle$ which can be expressed as

$$|B\rangle_e = \sqrt{\frac{e!(N - e)!}{N!}}(J_+)^e|G\rangle \tag{3.41}$$

with N the number of spins and $e = \{0 \ldots N\}$ the number of excitation in the spin ensemble. The set of superradiant spin states $|B\rangle_e$ form now the so called Dicke ladder. This ladder of equidistant states is climbed by increasing the number of excitations e in the spin ensemble (see Fig. 3.11). Where the degeneracy follows from Eq. 3.38. The number of possible spin states rises quickly as the number of excitations is increased. However, although the degeneracy rises always only one symmetrical spin state is found, but a rapidly increasing number of subradiant states with a maximum for $N/2$ excitations. The subsets of superradiant and subradiant states and their degeneracy for a given number of excitations e follow as

$$\frac{N!(2J + 1)}{(N/2 + J + 1)!(N/2 - J)!} \tag{3.42}$$

with $J = \sum_{k=0}^{e} |N/2 - k|$ spin manifolds. For the superradiant spin states with $k = 0$ a non-degenerate spin manifold is found, while the degeneracy and number of manifolds for subradiant spin states with $k > 0$ quickly grows with the number of excitations e. The spin state degeneracy in the Dicke model follows then as as:

- **Ground state:** One symmetrical spin state with $|J = \frac{N}{2}, m_J = -\frac{N}{2}\rangle = |G\rangle$
- **1st excited state** One symmetrical spin state with $|J = \frac{N}{2}, m_J = -\frac{N}{2} + 1\rangle = |G + 1\rangle$ and $N - 1$ subradiant states with $|J = \frac{N}{2} - 1, m_J = -\frac{N}{2} + 1\rangle$
- **2nd excited state** One super radiant state with $|J = \frac{N}{2}, m_J = -\frac{N}{2} + 2\rangle = |G + 2\rangle$ and $N - 1$ subradiant spin states with $|J = \frac{N}{2} - 1, m_J = -\frac{N}{2} + 1\rangle$ and $\frac{N(N-3)}{2}$ states with $|J = \frac{N}{2} - 2, m_J = -\frac{N}{2} + 2\rangle$

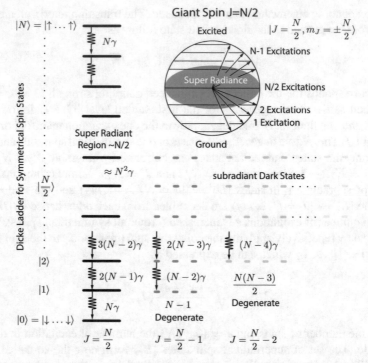

Fig. 3.11 The Dicke model for an ensemble consisting of N spins. A giant effective spin with $J = N/2$ features a ladder of symmetrical superradiant states for which the absorption or emission strength is collectively enhanced by $\sqrt{N}$ for the ground and excited states, respectively. If the giant spin is brought to a symmetrical spin state on the equator super radiance occurs since the single spin absorption or emission is increased by N

- ...
- **N excited state** One symmetrical spin state with $|J = \frac{N}{2}, m_J = \frac{N}{2}\rangle = |G + N\rangle$

In the frame work of the Dicke model a phenomenon called superradiance appears if the giant spin is brought to the equator on a symmetrical spin state. With other words if the number of excitations in the ensemble is equal to $N/2$ and the spins are in a symmetric superposition the total matrix element is increased since

$$J_+|B\rangle_e = \sqrt{(N-e)(1+e)}|B\rangle_e \tag{3.43}$$

with the selection rule $\Delta B_e = 1$. For a transition from the $N/2$ excited state to the $N/2 + 1$ excite state, i.e. the giant spin $J = N/2$ is on the equator, the transition matrix element has increased to

$$M_{\frac{N}{2} \to \frac{N}{2}+1} = \langle \frac{N}{2} + 1|J_+|\frac{N}{2}\rangle \approx N \tag{3.44}$$

and so the interaction of a cavity mode has increased from $\Omega = \sqrt{N}g$ to $\Omega = Ng$. Additionally the probability for spontaneous decay of any excited spin state according to Fermi's golden rule is proportional to the squared transition matrix element. For a spin ensemble in the absence of a cavity moded the spontaneous emission rate therefore drastically increases $\Gamma_{max} \approx \gamma N^2$. With a spin ensemble coupled to a resonator in the bad cavity regime $\kappa \gg \frac{g^2}{\gamma} \gg \gamma$ [6] the Purcell [7] enhanced spontaneous decay rate reads [8, 9]

$$\Gamma_{max} = \frac{N^2 g^2 \kappa}{\Delta^2 + (\kappa/2)^2} \tag{3.45}$$

with a spin cavity detuning Δ and cavity disspation rate κ. Hence the maximal decay rate on resonance is given by $\Gamma_{max} = 4N^2 g^2/\kappa$ emitting a superradiant pulse when the ensemble is prepared in the excited state. In contrast an uncorrelated and excited ensemble would just decay exponentially, whereas superradiance causes a strong intensity pulse with a delay time τ. This phenomenon was first described by R.H. Dicke in 1954 as superradiance and means that an excited ensemble of spins will decay much faster than a single spin, on a timescale given by $\tau \approx \frac{\kappa}{4Ng^2} = \frac{\tau_{single}}{N}$. One should note that the spin dissipation rate γ was replaced by $4g^2/\kappa$ which is a good approximation in the bad cavity limit and if the cavity field is the most prominent decay channel for an individual spin. If the cavity spin system is in the strong coupling regime this phenomenon gives also rise to strong non-linear effects when the photon number in the cavity is large compared to the number of spins $\langle a^\dagger a \rangle \to N/2$.

3.2.4 Low Excitation Limit

In the case of having a large spin ensemble coupled to a cavity mode containing a low number of excitations one can treat these excitations as non-interacting quasi particles. Such an approximation is known as Holstein-Primakoff transformation [10–12] which is possible if the number of photon $a^\dagger a = N$ is small compared to the number of spins N in the cavity. In an intuitive picture a large spin ensemble excited by a single exctiation is shared equally by all spins. Therefore the total energy is not sufficient to saturate a single spin nor the ensemble and effectively all spins remain in their ground state and only scatter the cavity field mode. If this is true one can replace a single anharmonic spin of the ensemble by a harmonic oscillator since the energy is not sufficient to go beyond the first Fock state of these harmonic oscillators. This is equivalent to a bosonization of the spin ensemble in which one maps the spin operators to bosonic creation and annihilation operators as

$$J_+ = b^\dagger \sqrt{1 - b^\dagger b} \text{ and } J_z = 2b^\dagger b - 1 . \tag{3.46}$$

Further the Hamiltonian in Eq. 3.36 can be modified treating spins and cavity as harmonic oscillators and subsequently reads

$$\mathcal{H} = \hbar\omega_c a^\dagger a + \hbar\omega_s b^\dagger b + \hbar\Omega (ab^\dagger + a^\dagger b) \tag{3.47}$$

with the assumption $J_+ \sim b^\dagger = \frac{1}{\sqrt{N}} \sum_j^N b_j^\dagger$, which is valid for spin states with low and high $m_s \sim \pm\frac{N}{2}$ values. This approximation is valid if the number of excitations in the system is smaller than approximately $\sqrt{N}$ or $b^\dagger b + a^\dagger a \ll 1$ from which $J_z \approx -1$ follows. As a consequence, in this regime the coupled spin ensemble does not leave the single excitation manifold of the Dicke model. However, if the number of photons in the cavity is increased the coupled system leaves the linear regime and due to the increased Rabi frequency strong non-linear effects are observable.

3.3 Coupling to Inhomogeneous Spectral Broadened Spin Ensembles

As shown in the previous section the Tavis-Cummings and Dicke Model is suitable to describe a spin ensemble consisting of N spins. In the idealized Dicke model spins or atoms would have identical transitions frequencies ω_s, but in order to include inhomogeneous spectral broadening one has to modify the corresponding Hamiltonian as

$$\mathcal{H} = \frac{\hbar}{2} \sum_j^N \omega_j \sigma_j^z . \tag{3.48}$$

The spin transitions frequency ω_j of the $j th$ spin, centered around a central spin frequency ω_s, describe now an ensemble of N spins which may be spectraly broadened. The variation of spectral frequencies is described by a spectral density of states $\rho(\omega)$ which is in the most general case a sum of many natural broadened lines. The natural line width follows from the Heisenberg uncertainty principle

$$(E_e - E_g)\Delta t = \hbar \tag{3.49}$$

relating any excited two level system to a finite life time. From this follows a full-width at half-max line width of a single spin or two-level system as

$$\gamma \geq \frac{1}{\Delta t} = (E_e - E_g)/\hbar . \tag{3.50}$$

Of course a spin line width truly limited by the Heisenberg uncertainty principle is a rather idealized situation. However the natural life time broadened spectral line shape

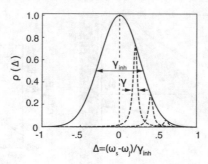

Fig. 3.12 Inhomogeneously broadened ensemble of spins. Spins with transition frequency ω_j and a Lorentzian natural lineshape with width γ centered around a central frequency ω_s. If γ is small compared to the variation of ω_j the resulting envelope and line shape of the ensemble will be Gaussian characterized by a width γ_{inh}

of such a transition is $\rho(\omega)_{hom}$ and can be approximated by a Lorentzian function of width γ.

If for the spectral spin frequencies ω_j a Maxwell-Boltzmann distribution is assumed, the resulting spectral line shape of $\rho(\omega)$ is a convolution

$$\rho(\omega) = \int_{-\infty}^{\infty} \rho(x)_{inh}\rho(\omega - x)_{hom}dx \qquad (3.51)$$

which results in inhomogeneous spin distribution $\rho(\omega)_{inh}$ of width γ_{inh} (Fig. 3.12). Three different regimes can be distinguished:

- If the inhomogeneous line width is large compared to the natural line width of a single spin the ensemble absorption or spectral density of states will follow a Gaussian distribution $\rho(\omega) \approx \rho(\omega)_{inh}$.
- If the inhomogeneous line width is small compared to the natural life time of a single spin the density of states follows a Lorentzian distribution $\rho(\omega) \approx \rho(\omega)_{hom}$.
- In real systems one often finds an intermediate regime where the spectral line shape is neither of pure Lorentzian nor Gaussian nature but rather a convolution of both distributions known as Voigt profile or q-Gaussian.

3.3.1 Equidistant Discretized Spin Ensembles

In the case of having an ensemble of spins in the cavity mode volume and assuming a homogeneous cavity spin interaction strength g the collectively enhanced coupling strength scales as $\Omega = \sqrt{N}g$. If the spin transition frequencies are Gaussian distributed with $\gamma_{inh} = \Omega$ one finds two polariton modes $|\pm\rangle$ and a bath of subradiant state $|D\rangle$ after diagonalizing a Tavis-Cummings Hamiltonian. In Fig. 3.13 the derived eigenenergy spectrum is shown for thirty spins with homogeneous single spin Rabi

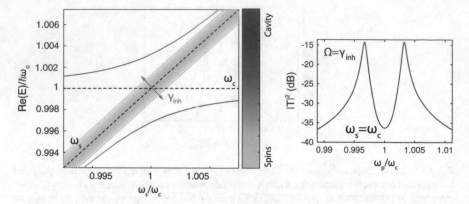

Fig. 3.13 Eigenenergy spectrum of thirty spins with Gaussian distributed transition frequencies centered around ω_s. The collective enhanced coupling strength Ω is assumed to be equal to the full-width at half-max spectral density γ_{inh}

frequencies and $\gamma_{\text{inh}} = \Omega$. The inhomogeneous broadening gives rise to a lifted degeneracy and the spectral density of the subradient states follows $\rho(\omega)$.

Such an inhomogeneous line broadening will cause dephasing and increases the total decay rate Γ of the system. An intuitive picture how this accelerates the evolution of a superradiant spin state

$$|B\rangle = \frac{1}{\sqrt{N}}(|\uparrow\rangle_1 + |\uparrow\rangle_2 + \cdots + |\uparrow\rangle_N) \tag{3.52}$$

into a subradiant spin state can be made by time evolution of $|B\rangle$. Therefore one has to account for each single spin frequency and thus write

$$|B(t)\rangle = \frac{1}{\sqrt{N}}(|\uparrow\rangle_1 e^{-i(\omega_s-\omega_1)t} + |\uparrow\rangle_2 e^{-i(\omega_s-\omega_2)t} + \cdots + |\uparrow\rangle_N e^{-i(\omega_s-\omega_N)t}) . \tag{3.53}$$

Such a state will de-phase and evolve into a subradiant state $|D\rangle$ as all spins pick up a relative phase shift to each other. The time constant on which this process occurs is known as free induction decay in electron or nuclear spin resonance and given by the width of the inhomogeneous broadened spin ensemble

$$T_2^* = \frac{1}{\gamma_{\text{inh}}} . \tag{3.54}$$

This process is reversible by applying a refocusing pulse, which can be understood as a time reversal and results in a spin echo and is limited by the single spin dissipation rate γ, as will be shown in the last experimental Chapter of this thesis.

As discussed, one can introduce the spectral variation of spin frequencies in an ensemble of N spins and describe the coupled system by an interaction Hamiltonian as introduced earlier but extended to $N + 1$ dimensions

$$\mathcal{H} = \hbar \begin{pmatrix} \omega_c + i\kappa & -ig & -ig & \dots & ig \\ ig & \omega_{s1} + i\gamma & 0 & \dots & 0 \\ ig & 0 & \omega_{s2} + i\gamma & \dots & 0 \\ \vdots & \vdots & \vdots & \ddots & \vdots \\ ig & 0 & 0 & \dots & \omega_{sN} + i\gamma \end{pmatrix} \tag{3.55}$$

from which the system transmission follows

$$T = \text{Tr}\{\text{inv}(\omega_p \mathbf{1} - \mathcal{H})\}\,|_{ij=11}\,. \tag{3.56}$$

when probed by a driving field with frequency ω_p which is shown in Fig. 3.13.

3.3.2 Non-equidistant Discretized Spin Ensembles

Holding the cavity spin coupling constant and changing the individual spin transition frequency is the most natural way of describing such a large spin ensemble. Nevertheless in a different approach one modifies the spectral coupling density g_μ according to the spin ensemble line shape $\rho(\omega)$. Such a step is justified if the number of spins is very large and the single spin coupling strength $g \ll \Omega$. Therefore in a spin ensemble of N spins the spectral ensemble profile reads

$$\rho(\omega) = \frac{\sum_j^N g_j^2 \delta(\omega - \omega_j)}{\Omega^2} \rightarrow g_\mu = \Omega \sqrt{\rho(\omega_\mu) / \sum_l \rho(\omega_l)} \tag{3.57}$$

with a collective coupling strength $\Omega = \sqrt{\sum_j^N |g_j|^2}$. Similar to the previous section a Hamiltonian can be established reading

$$\mathcal{H} = \hbar \begin{pmatrix} \omega_c + i\kappa & -ig_1 & -ig_2 & \dots & -ig_N \\ ig_1 & \omega_1 + i\gamma & 0 & \dots & 0 \\ ig_2 & 0 & \omega_2 + i\gamma & \dots & 0 \\ \vdots & \vdots & \vdots & \ddots & \vdots \\ ig_N & 0 & 0 & \dots & \omega_N + i\gamma \end{pmatrix} \tag{3.58}$$

with a spin coupling density corresponding to the spectral line shape following from Eq. 3.57. The spin components in Eq. 3.58 can now considered to be spin packets rather than single spins. Therefore g_μ represents the spectral coupling density of spin packets which is related to spectral spin density. Hence the spin ensemble can be discretized with the knowledge of $\rho(\omega)$ (Fig. 3.14). One should take great care in

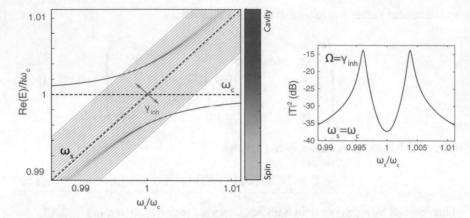

Fig. 3.14 Eigenenergy spectrum for Gaussian spectral coupling density profile g_μ. The collective enhanced coupling strength Ω is assumed to be equal to the full width half max spectral density γ_{inh}. We observe to polariton modes residing in a bath of dark states

choosing a single spin dissipation rate γ. If then umber of spin packets is finite one hast to choose γ in such a way that there is an effective energetic overlap between individual spin packets. As the density increases γ will converges to its true value but can cause an excessive numerical overhead in calculations.

References

1. S. Haroche, J.-M. Raimond, *Exploring the Quantum: Atoms, Cavities, and Photons* (OUP Oxford, April 2013). ISBN 978-0-19-968031-3
2. M.O. Scully, M.S. Zubairy, *Quantum Optics* (Cambridge University Press, September 1997). ISBN 978-0-521-43595-6
3. D.F. Walls, G.J. Milburn, *Quantum Optics* (Springer Science & Business Media, January 2008). ISBN 978-3-540-28573-1
4. M. Tavis, F.W. Cummings, Exact solution for an n-molecule-radiation-field hamiltonian. Phys. Rev. **170**(2), 379–384 (1968)
5. R.H. Dicke, Coherence in spontaneous radiation processes. Phys. Rev. **93**(1), 99–110 (1954)
6. A. Predojević, M.W. Mitchell, *Engineering the Atom-Photon Interaction: Controlling Fundamental Processes with Photons, Atoms and Solids.* (Springer, July 2015). ISBN 978-3-319-19231-4
7. E.M. Purcell, Proceedings of the american physical society. Phys. Rev. **69**(11–12), 681 (1946)
8. H. Iwase, D. Englund, J. Vucković. Analysis of the purcell effect in photonic and plasmonic crystals with losses. Opt. Express. **18**(16):16546–16560, (August 2010). ISSN 1094-4087
9. J.A. Mlynek, A.A. Abdumalikov, C. Eichler, A. Wallraff, Observation of Dicke superradiance for two artificial atoms in a cavity with high decay rate. Nat. Commun. **5**, 5186 (2014)
10. H. Primakoff, T. Holstein, Many-body interactions in atomic and nuclear systems. Phys. Rev. **55**(12), 1218–1234 (1939)

11. Z. Kurucz, J.H. Wesenberg, K. Mølmer, Spectroscopic properties of inhomogeneously broad-
    ened spin ensembles in a cavity. Phys. Rev. A **83**(5), 053852 (2011)
12. D.O. Krimer, S. Putz, J. Majer, S. Rotter, Non-Markovian dynamics of a single-mode cavity
    strongly coupled to an inhomogeneously broadened spin ensemble. Phys. Rev. A **90**(4), 043852
    (2014)

Chapter 4
Experimental Implementation—Solid-State Hybrid Quantum System

In this chapter I will introduce the devices implemented to carry out the presented experiments. The hybrid quantum system consists of a superconducting microwave cavity coupled to macroscopic ensemble of nitrogen-vacancy (NV) defect centers in diamond. Therefore the first sections will briefly introduce transmission line resonators and NV centers. In the latter the cryogenic environment in which the actual experiments are preformed is shown. Followed by a discussion of the employed spectroscopic techniques. The main part of the discussed experiment was setup and developed in work performed earlier [1–3] and builds up on these efforts.

4.1 Micro-Wave Cavities

In this section I will briefly show, how the concepts discussed in Chap. 2 can be realized and employed in order to study electrical oscillators in the microwave (MW) frequency domain. Superconducing microwave cavities have been a versatile tool in experimental physics (e.g. ultra sensitive detectors [4]) and have become a work horse in circuit cavity QED as a "quantum bus" interface [1, 2, 5, 6, 6–9].

4.1.1 Distributed Electrical Resonators

A straight forward realization of a MW cavity is a coplanar transmission line of length l. At each side of the waveguide, the electromagnetic field is reflected and a standing wave matching $\omega = kc$ is created. In Sect. 2.1.2 it was shown that the cavity creates boundary conditions and only modes which are an integer multiples n of $k = \pi/l$ are allowed. The fundamental resonance is given by $n = 1$, while resonance with $n > 1$ are higher harmonics of the cavity. One should note that the Q factor of the cavity scales with n for higher harmonics. However, in the latter only on the fundamental resonance with $n = 1$ is considered. A one dimensional version

© Springer International Publishing AG 2017

S. Putz, *Circuit Cavity QED with Macroscopic Solid-State Spin Ensembles*,
Springer Theses, DOI 10.1007/978-3-319-66447-7_4

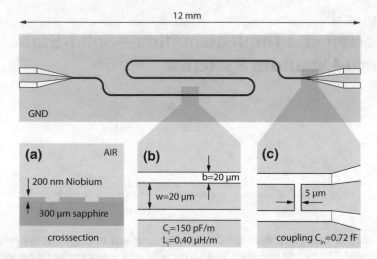

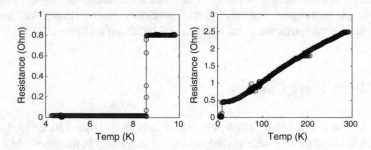

Fig. 4.1 Design of a coplanar waveguide (CPW) resonator. **a** A 200 nm thick niobium film is deposited on a sapphire substrate. **b** By optical lithography a central conductor, with a width of $w = 20 \, \mu m$ and separated by two gaps $b = 8.3 \, \mu m$ is created. (**c**) This transmission line is interrupted by two capacitors at each end which create a $\lambda/2$ MW cavity

Fig. 4.2 Four point measurement of the critical temperature $T_C = 8.5$ K and a residual-resistivity ratio $RRR = 5.533$ of a 200 nm thick Niobium film

of a transmission line is a coplanar waveguide resonator (CPW) and the design used in this thesis is shown in Fig. 4.1.

The microwave cavities used to carry out the presented experiments are made out of a 200 nm niobium superconducting thin film sputtered on a 300 μm thick sapphire substrate. Niobium is a type II superconductor and a measurement of the critical temperature (T_C) is presented in Fig. 4.2. The high quality of the deposited thin film allows a high $T_C = 8.5$ K (in bulk 9.2 K) and a residual-resistivity-ratio $RRR = 5.533$. Such a thin film is then structured by optical lithography and a transmission line, intercepted by two coupling capacitors, is created. A realization of a waveguide with $50 \, \Omega$ characteristic impedance is a central conductor of with $w = 20 \, \mu m$ separated by two gaps $b = 8.3 \, \mu m$ from the ground plane and 200 nm thickness. From this geometry a unit inductance $L_l \sim 0.4 \, \mu H/m$ and capacitance

C_I ~150 pF/m per unit length [2] follow from finite element simulations and are known as the distributed parameters. From the simulated values of C_I and L_I the characteristic impedance is estimated to be $Z_0 = \sqrt{L_I/C_I} \approx 50\,\Omega$.

The resonator frequency is determined by the separation of the two coupling capacitors, which defines the length l of the cavity. Therefore the total capacitance and inductance follow as $C = C_I l$ and $L_I l$ respectively. The cavity frequency and fundamental resonance is then given by $\omega_c = 1/\sqrt{LC}$. The design shown in Fig. 4.1 has a length of $l = 21.3\,\text{mm}$ and the resonance frequency is estimated as follows. For an ideal transmission line the phase velocity is given by

$$v_p = \frac{1}{\sqrt{L_I C_I}} \tag{4.1}$$

with the dispersion relation

$$\omega_c = k v_p = \frac{\pi}{l} \frac{1}{\sqrt{L_I C_I}} \tag{4.2}$$

which determines the resonance frequency for the fundamental resonance of the $\lambda/2$ cavity. For the given C_I and L_I the cavity fundamental resonance is estimated to be $\omega_c/2\pi \approx 3\,\text{GHz}$ omitting the frequency pull due to the coupling capacitors C_K. The coupling to an external feed line will lower the cavity frequency by approximately $-\omega C_K/C$ [9].

The spectral transmission through the cavity (scattering parameter $|S_{21}|^2$) of the above discussed design is shown in Fig. 4.3. From a Lorentzian fit of the experimental data a resonance frequency $\omega_c/2\pi = 2.69\,\text{GHz}$ and a cavity line width of $\kappa/2\pi = 462\pm2\,\text{kHz}$ (HWHM) with a quality factor of $Q = 2910$ is determined. The effective electric permittivity of the cavity follows from the relation of frequency and phase velocity reading

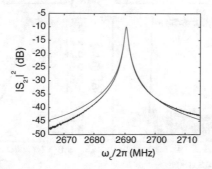

Fig. 4.3 Transmission measurement of CPW superconducting MW cavity. A Lorentzian fit gives a resonance frequency $\omega_c/2\pi = 2.69\,\text{GHz}$ and a cavity line width of $\kappa/2\pi = 462\pm2\,\text{kHz}$ (HWHM) with a quality factor of $Q = 2910$ in the high power limit

$$\omega_c \frac{l}{\pi} = \frac{1}{\sqrt{\epsilon_r \epsilon_0 \mu_0}} . \tag{4.3}$$

The estimated $\epsilon_r \sim 7$ is in good agreement with the theoretical values for the relative permittivity of the sapphire substrate $\epsilon_r =9$–11 and an additional dielectric diamond loading with $\epsilon_r =5$–10. The quality factor of the empty cavity $Q_{\text{empty}} =40{,}560$ and $\omega_c/2\pi = 2.89$ GHz is measured for comparison when the crystal was removed. The dielectric loading causes a frequency shift $\Delta/2\pi = 200$ MHz and the empty over-coupled cavity becomes effectively under-coupled with $g_c \sim 0.08$. The time dependent transmitted intensity through the cavity $|A|^2$ is measured by a autodyne detection scheme. As is shown in Fig. 4.4 during a $8\,\mu s$ long and resonant driving pulse ($\omega_c = \omega_p$) the cavity transmission builds up and reaches a steady state value. After the drive is switched of the transmitted intensity decays with a characteristic time constant given by $\tau = \frac{1}{2\kappa}$. An exponential fit to the decaying signal with $e^{-t/\tau}$ shows good agreement with the measured data and κ derived from the scattering parameter $|S_{21}|^2$.

The linewidth κ can be interpreted as the photon lifetime in the resonator. Therefore, this dissipation rate determines the average photon number in the cavity at a given input power P_{in}. For example the scattering parameter $|S_{21}|^2$ shown in Fig. 4.3 is measured with an intensity $P_{\text{in}} = -100\,\text{dBm} = 0.1\,\text{pW}$ at the cavity input port. This stimulus induces an estimated voltage of $V \approx 4.8 \times 10^{-4}\,\text{V}$ and a current of $I \approx 9.7 \times 10^{-6}\,\text{Ampere}$ in the LC circuit. The mean photon number $|A|^2 = |\alpha|^2 = \langle N \rangle$ for this drive power can be estimated as $P_{\text{in}}/\hbar\omega_c\kappa$, corresponding to $\sim 1.25 \times 10^5$ photons on average stored in the cavity.

The field distribution of a coplanar waveguide can be calculated by using conformal mapping techniques. Since the formulas are rather lengthy and not very intuitive only the result of such a calculation is shown and follows [2, 10]. In Fig. 4.5 the magnetic field distribution normalized to the vacuum energy is plotted. The field

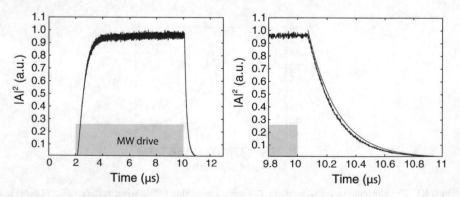

Fig. 4.4 Time domain measurement of a CPW resonator. During a $8\,\mu s$ long rectangular and resonant driving pulse ($\omega_c = \omega_p$), the signal builds *up* and reaches a stationary state. When the drive is switched of the transmitted intensity decays with a characteristic time constant given by $\kappa/\pi = 924 \pm 2$ kHz (FWHM) indicated by an exponential fit *(red line)*

Fig. 4.5 Simulation of the magnetic field produced by a CPW normalized to the vacuum field. The field strength at a distance of 1 μm is typically of the order of 10 μG corresponding to a single spin Rabi frequency of $g \approx 10\,$Hz

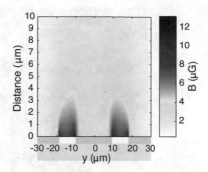

strength reaches its maximal values over the gaps between conductor and ground plane and is at a distance of 1 μm typically of the order of 10 μG. This vacuum field strength accounts for the single spin vacuum Rabi frequency. With the electron magnetic moment of $\mu_e = 1.4\,$MHz/G this corresponds then to a typical single spin cavity interaction strength $g/2\pi \approx 10\,$Hz.

4.1.2   Lumped Electrical Resonators

In the previous section the microwave cavity was characterized in terms of distributed parameters. This means that the length l of the structure determines the resonator wavelength. A different approach to realize an electrical LC circuit is to directly fabricate an inductance and a capacitance in a parallel configuration. For such structures the wavelength can become much larger than the actual size of the device. Therefore this circuit is called a lumped element resonator (LER). As a consequence the resonator has only one fundamental resonance, since the transmission line model, with possible higher harmonics, breaks down.

A possible realization of such a parallel meander inductor L and an inter-digit capacitor C capacitively coupled to a 50 Ω feed line is shown in Fig. 4.6. A finite element simulation of $L = 2.7\,$nH and $C = 0.43\,$pF allows to make a course estimate on the resonance frequency. It is then useful to simulate the complete structure and determine cavity frequency ω_c and Q factor. One of the big advantages is that the lateral dimensions can be varied and the size and shape of the cavity mode volume is controllable to some extend. It is further possible to increase the magnetic field strength produced by the cavity, by decreasing the inductance L hence unbalancing the circuit.

The measured transmission of the discussed design is shown in Fig. 4.7. The resonance frequency $\omega_c/2\pi = 3.929\,$GHz and cavity line width of $\kappa/2\pi = 150 \pm 1\,$kHz (HWHM) with a quality factor of $Q = 13036$ are estimated by a Lorentzian fit. The cavity response in the time domain is shown in Fig. 4.8. During the 8 μs long and resonant ($\omega_c = \omega_p$) driving pulse, the cavity field builds up and reaches

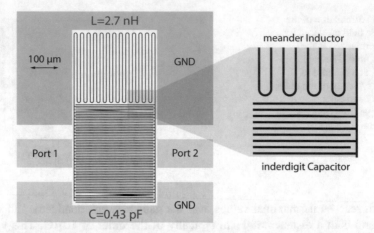

Fig. 4.6 Design of a transmission lumped element (TLER) resonator. The parallel inductance L and capacitance C are capacitively coupled in transmission to a $50\,\Omega$ *feed line*

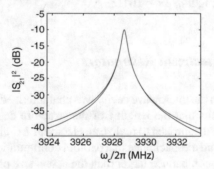

Fig. 4.7 Transmission measurement of a TLER microwave resonator. A Lorentzian fit gives a resonance frequency $\omega_c/2\pi = 3.929\,\text{GHz}$ and a cavity line width of $\kappa/2\pi = 150 \pm 2\,\text{kHz}$ (HWHM) with a quality factor of $Q = 13036$ in the high power limit

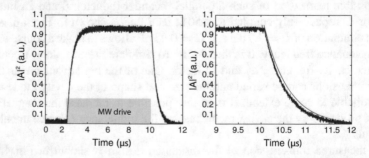

Fig. 4.8 Time domain measurement of a TLER resonator. After the resonant drive is switched off the transmitted intensity decays with a characteristic time constant given by $\tau = \frac{1}{2\kappa}$ with $\kappa/2\pi = 150 \pm 2\,\text{kHz}$

its steady state intensity. When the drive is switched off the cavity amplitude decays with a characteristic time constant given by $\tau = \frac{1}{2\kappa}$. As demonstrated TLER and CPW resonators show an exactly similar cavity responses.

4.2 Nitrogen Vacancy Center Spin Ensembles

In the theoretical part of this thesis the interaction of electron spins with the electromagnetic field inside a cavity was discussed. Different para-magnetic defects in semiconductor solid-state crystals have been used to realized strong cavity spin interaction [11–16]. However under these defects, one of the most promising impurities is the negatively charged nitrogen-vacancy (NV) color center in diamond [17–20]. A NV defect center center consists of a substitutional ^{14}N nitrogen atom and an adjacent lattice vacancy, as is shown in Fig. 4.9. Different charge configurations of the defect center are possible, but hereafter only the negatively charged NV center is considered. The electronic structure of a NV center follows from, two non-bounded valence electrons from the nitrogen atom, and three unpaired electrons from the vacancy. If an additional sixth electron is trapped by the defect this results in total in two unpaired electrons and a negative charge state. This configuration forms the so called negatively charged nitrogen-vacancy (NV) center, which is an electron spin triplet with spin quantum number $S = 1$. The defect features a C_{3v} symmetry and the energy level scheme is shown in Fig. 4.9 indicating the characteristic zero-phonon-line (ZPL) at 637 nm. The optical transition between $m_s = 0$ states

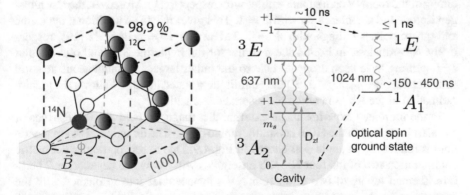

Fig. 4.9 *Left* Illustration of a nitrogen-vacancy (NV) defect center in diamond. A substitutional ^{14}N nitrogen atom and an adjacent lattice vacancy create a molecule like defect center. Due to the four different orientations of an adjacent lattice vacancy four distinct NV sub ensembles are possible in the diamond crystal. Zeeman tuning of the NV spin resonance transitions allows addressing two or all four NV sub ensembles by controlling the magnetic field angle ϕ in the (100) crystal plane. *Right* Energy level scheme of a NV center. A zero-phonon *line* (ZPL) at 637 nm allows optical readout of the *color* center. However, the microwave cavity interacts with the electron spin triplet in the optical ground state

is spin conserving, whereas $m_s = \pm 1$ states have a high probability to decay into metastable levels by a non-radiative process. This is one of the key properties of NV centers, which allows to optically read out the electron m_s spin state. However, in the presented experiments the defect is never optically excited. The cavity interacts with the electron spin triplet in the optical ground state only.

4.2.1 Nitrogen Vacancy Center Level Structure

The two unpaired electrons of the NV center give rise to a zerofield splitting term $\mathbf{D}_{zf}$. This splitting is due to a dipolar spin-spin interaction which lifts the degeneracy of the spin triplet at zero external field. The zerofield splitting term determines the spin quantization axis with respect to the NV axis. The NV axis is defined by the orientation of the substitutional nitrogen atom and the adjacent lattice vacancy. In the face-cubic-centered diamond crystal four different orientations are possible for an adjacent lattice vacancy. Therefore four different orientations of $\mathbf{D}_{zf} = (D_x, D_y, D_z)$ are possible, giving rise to four NV spin species in the crystal. The spin Hamiltonian describing the NV spin triplet $S = (S_x, S_y, S_z)$ in an external magnetic field $\mathbf{B}_{ext}$ reads then for one spin species

$$\mathcal{H}_{NV} = \hbar D_{zf} S_z^2 + \hbar E_{zf}(S_x^2 - S_y^2) + \hbar \mu_{NV} \mathbf{B}_{ext} S \tag{4.4}$$

with zerofield spitting terms $D_{zf} = \frac{3}{2} D_{zf} = 2.88\,\text{GHz}$ and $E_{zf} = \frac{D_x - D_y}{2} = 8\,\text{MHz}$ and a spin magnetic moment $\mu_{NV} = 28\,\text{MHz/mT}$. The Zeeman term allows magnetic tuning of the four NV spin subensembles with respect to $\mathbf{D}_{zf}$. Moreover, the ^{14}N nitrogen isotope has a nuclear spin with $I = 1$. This gives rise to an additional hyperfine and quadrupolar splittings with $\hat{A} = 2.2\,\text{MHz}$ and $\hat{P} = 5.04\,\text{MHz}$ [19], respectively. Nevertheless, in Eq. 4.4 these terms for the hyperfine $S\hat{A}I$ and quadrupolar $I\hat{P}I$ splitting have been dropped. Due to the rather larger inhomogeneous spectral spin broadening ($\gamma_{inh} =\gg \hat{A}, \hat{P}$) the Hamiltonian stated in Eq. 4.4 allows the precise estimation of the NV spin ensemble resonance transitions.

In the presented experiments an external d.c. magnetic field $\vec{B}_{ext} = B_{ext}(\cos \phi, \sin \phi, 0)$ is applied in the (100) crystallographic plane. The direction of the magnetic field is rotated by an angle ϕ as shown in Fig. 4.9. Due to this particular geometric configuration two of the four NV sub ensembles will always be degenerate. Therefore Zeeman tuning of two or all four NV subensembles into resonance with the microwave cavity is possible, when the external magnetic field is applied in the (100) plane as is shown in Fig. 4.10.

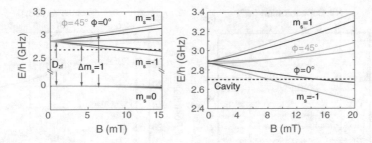

Fig. 4.10 *Left* NV energy level scheme derived from the NV spin Hamiltonian in Eq. 4.4. The zerofiled splitting $D_{zf} = 2.88$ GHz lifts the degeneracy and only moderate magnetic field strengths are needed to tune the spin energy levels into resonance with the cavity *(black dotted line)*. *Right* Spin and cavity resonance transitions derived form the NV Hamiltonian. As the magnetic field angle ϕ is rotated NV subsensembles become non degenerate

4.2.2 Thermal Polarization and Spin-Spin Interactions

Only moderate magnetic field strengths are necessary to Zeeman tune the NV spins into resonance with a microwave resonator due to the large zerofield splitting. This makes NV centers suitable for the combination especially with superconducting qubits [21] since flux focusing effects already at low magnetic field strengths can degrade superconducting devices. Additionally the thermal polarization of the spin ensemble is possible at zero external magnetic fields and at moderate temperatures achievable in a standard dilution refrigerator. The large zerofield splitting parameter can be mapped to a temperature

$$T_{zf} = \frac{h D_{zf}}{k_B} \approx 138 \text{ mK} \tag{4.5}$$

with the Boltzmann constant k_B. Further the thermal polarization rate of the NV spin triplet is estimated by a Boltzmann statistics

$$p_{m_s} = \frac{e^{\frac{-\mathcal{E}_{m_s}}{k_B T}}}{\sum e^{\frac{-\mathcal{E}_{m_s}}{k_B T}}} \tag{4.6}$$

with the NV eigenenergies $\mathcal{E}_{m_s}$ following from Eq. 4.4. The result for zero external magnetic field is shown in Fig. 4.11. The presented experiments are typically performed at temperatures of ≤ 25 mK at which the NV spin polarization rate is estimated to be greater than 99.9%.

Up to now only single spins have been considered to estimate the spin transitions frequencies and the polarization rates. However, a macroscopic spin ensembles will be employed to achieve strong spin cavity interactions. The theoretical discussion in Chap. 3 assumed independent spin two-level systems. Therefore to estimate if

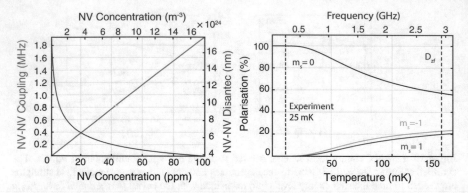

Fig. 4.11 *Left* Interaction strength of two NV spin dipoles as a function of the concentration and mean distance in the diamond crystal. *Right* NV spin triplet polarization as a function of temperature, the experimental temperature is considerably smaller than the NV zerofield splitting and ensures thermal spin polarization rates close to unity

this treatment is accurate the NV dipole spin-spin interaction strengths have to be negligibly small. The NV spin-spin interaction scales with the NV spin concentration of the ensemble. The concentration of carbon atoms in diamond corresponds to $c_d = 1.76 \times 10^{29} \frac{^{12}C}{m^3}$ with the diamond lattice constant $a_d = 0.357$ nm. If a NV concentration of one part per million (ppm) is assumed this corresponds to a mean spin-spin distance of

$$r = \sqrt[3]{\frac{1}{c_d \times 10^{-6}}} = 17.85 \text{ nm} . \tag{4.7}$$

The resulting NV dipole-dipole interaction strength of two adjacent NV spins in the ensemble is then on average estimated to be

$$\frac{h\mu_0\mu_{NV}^2}{4\pi r^3} \approx 10 \text{ kHz} \tag{4.8}$$

in a diamond crystal containing one ppm of NV centers. This interaction scales linearly with the NV concentration which is shown in Fig. 4.11. This shows that the estimated spin-spin interaction is over three orders of magnitude greater than the single spin Rabi frequency. However it is still rather small compared to the line width of the cavity and broadened spin ensemble and therefore is neglected in the discussion presented later on.

In a macroscopic spin ensemble every individual NV spin will have a slightly different transition frequency which gives rise to spectral line broadening. Different effects can be identified in causing inhomogeneous spin broadneing. One major source is the spin bath [22–26] created by excess electron and nuclear spins in the host material. Natural diamond crystals consist of 98.9% spin less ^{13}C atoms and have a natural abundance of 1.1% ^{13}C carbon isotopes with nuclear spin $I = 1/2$. Additional excess nitrogen impurities P1 centers with electron spin $S = 1/2$ act as the main source of decohrence and dephasing in the spin host material. In dilute crystals the bath of ^{13}C nuclear spins will be the dominant source of dephasing with an estimated NV to ^{13}C spin coupling strength of ≈ 40 kHz. Electron spin densities (NV or P1) of less then one ppm equal the influence of the ^{13}C nuclear spin bath, due to their over a factor of thousand larger gyro magnetic ratio. Additionally spin dephasing and dissipation is introduced by crystallographic lattice stress. Lattice deformations due to irradiation with electrons or neutrons causes a variation of the zerofield splitting parameter of every single NV spin in the ensemble.

In the experiments presented a diamond crystal with natural abundance of ^{13}C carbon isotopes and 100–200 ppm P1 centers is used. A total density of ≈ 2 ppm NV centers is achieved by enhancing the type Ib high-pressure high-temperature diamond (HPHT) crystal. This is done by 50 h of neutron irradiation with a fluence of 5×10^{17} cm^{-2} and annealing the crystal for 3 h at 900 °C [27]. In this crystal the excess nitrogen P1 centres ($S = 1/2$) and additional lattice stress serve as main source of decoherence. Hence this contributions by far exceed dephasing caused by the natural abundant 1.1% ^{13}C spin bath. The spin ensemble is initially characterized by a room temperature confocal laser scanning microscope. Zerofield splitting parameters $D_{zf} = 2.88$ GHz and $E_{zf} = 8$ MHz and ensemble linewidth $\gamma_{inh}/2\pi \approx 10$ MHz have been estimated by optically detected magnetic resonance (ODMR) spectroscopic measurements prior to the performed experiments [3]. The mean NV spin-spin interaction strength in this ensemble is estimated to be on the order of 10 kHz, which justifies the assumption of treating the ensemble as non interacting spins at least to the first order.

4.3 Experimental Setup

The experiments presented in this thesis are carried out in a ^{3}He/^{4}He dilution refrigerator. The used Oxford instruments Triton 400 fridge has a cooling power of up to 400 μW at a temperature of 100 mK. The cryogen free refrigerator uses a pulse tube cooler to create temperatures of below 4–5 K. Additionally in a dilution unit a ^{3}He/^{4}He gas is circulated and passes heat exchangers and a mixing chamber. In this chamber, at temperatures below 1 K ^{3}He/^{4}He, the condensed gas will be separated in a concentrated and dilute ^{3}He phase. In the dilution process ^{3}He flows from the concentrated into the dilute phase by crossing a phase boundary. This process absorbs energy of the system and is the main cooling principle in a ^{3}He/^{4}He dilution refrigerator.

The actual experiment takes place at the base plate of the cryostat. During the discussed experiments the base temperature of the refrigerate was typically below 25 mK. Such low temperatures are necessary in order to thermally polarize the NV

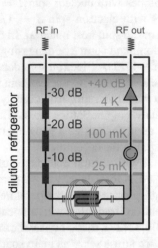

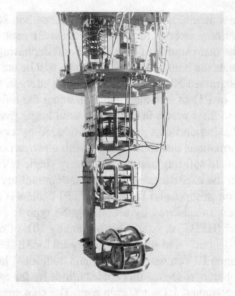

Fig. 4.12 *Left* Schematic drawing of the setup housed in a dilution refrigerator. *Right* Picture of the cryostat base plate with the attached experiments operated at 25 mK. The solid-state hybrid device is be placed in one of the three Helmholtz coil cages

Fig. 4.13 Picture of the superconducting chip bonded into a printed circuit board, and loaded with an enhanced diamond crystal

ensemble and to ensure high cavity Q values, as is discussed in the previous sections. The setup is then equipped with microwave cables and three-dimensional Helmholtz coil cages. These superconducting magnets allow to apply magnetic field strengths of up to 100 mT in an arbitrary direction. A schematic drawing of the fridge with different temperature plates is shown in Fig. 4.12, as well as a picture showing the fridge base plate with three three-dimensional Helmholtz coil cages attached to it.

The heart of the experiment is enclosed in one of these coil cages. Hosted in a copper box the superconducting chip is bonded into a printed circuit board (PCB). The resonator is then loaded by placing the enhanced diamond crystal on top of the planar cavity. An example of such a hybrid solid-state device is depicted in Fig. 4.13. The transmission line resonator is then connected to a waveguide on the PCB which can be connected by semi-rigid coaxial cables with SMA plugs. These lines are properly heat sinked and routed through the dilution refrigerator, which allows to measure the two-port device in transmission. The microwave line uses stainless steel cables, if a connection between two fridge stages with different temperatures is made. This ensures thermal isolation due to the low heat conductivity of stainless steel. Additional heat sinks of the coaxial line are crated by using microwave attenuators, which is important if measurements are performed on the single photon level. As a figure of merit the attenuation should be greater than the temperature ratio between to fridge plates. However the initial attenuation of 60 dB was removed, since the measurement intensities in the presented experiments were >1000 Photons in the cavity.

4.4 Measurement Scheme

In the performed experiments the hybrid device is stimulated and tested by either a fast autodyne detection setup or a vector network analyzer (VNA). Steady state measurement of the cavity scattering parameters (e.g. $|S_{21}|^2$) are rather straight forward with a VNA. However the time resolution is rather coarse and allows only to study very slow dynamical effects. Therefore a custom made autodyne detection setup is used to capture the system dynamics in the subnanosecond time domain. In combination with a fast arbitrary waveform generator microwave signals with only a few nanoseconds pulses duration and up to 500 mW intensity can be applied.

The autodyne detection scheme is shown in Fig. 4.14. A signal generator provides a Gigahertz microwave tone with an adjustable probe frequency ω_p. During a single shot, this frequency is fixed and the stimulus is split in two signal paths.

Fig. 4.14 Schematic drawing of the autodyne measurement scheme. A first frequency mixer upconverts two quadrature signals provided by an arbitrary waveform generator. The transmitted intensity is amplified and *down* converted by a second frequency mixer. Both quadrature signals are then recorded by a fast oscilloscope

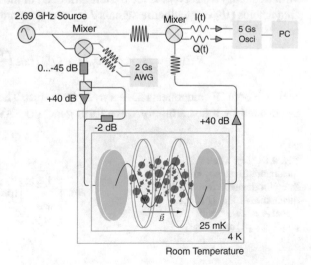

A signal component is routed through the experiment while the reaming one serves as a reference signal for the down conversion of the transmitted signal through the cavity. The stimulus is initially quadrature amplitude modulated by a microwave frequency mixer and a fast arbitrary waveform generator (AWG). This modulation allows to control the amplitude, length and phase of the microwave pulse fed towards the experiments. After this pulse shaping the signal can be dynamically attenuated $(0...-45 \text{ dB})$ and routed through a high power amplifier with $+40$ dB gain. After the pulse shape and intensity has been adjusted the signal enters the cryostat and is attenuated by -2 dB to thermalize the microwave coaxial cable at 4 K. The device under test is then measured in transmission and the signal from the cavity output is fed into a low temperature cryogenic amplifier on the 4 K stage. This amplifier with a noise temperature of 4 K has a gain of $+45$ dB and the amplified signal is then routed back to room temperature outside of the fridge. The signal transmitted through the cavity is then combined with the reference signal by a second frequency mixer. The down converted quadratures are low pass filtered and after an additional amplification recorded by a fast 4 GS/s oscilloscope. From the quadratures $I(t)$ and $Q(t)$ the oscillator amplitude $|A(t)| = \sqrt{I(t)^2 + Q(t)^2}$ and phase $\phi(t) = \arctan(Q(t)/I(t))$ is calculated. Finally the measured signal is then identified with the expectation value of the cavity photon number $|A|^2 = \langle a^\dagger a \rangle$.

4.4.1 Up and Down Conversion

In the discussed measurement scheme two quadrature frequency mixers are used. The mixer is a four port device and is schematically drawn in Fig. 4.15. The frequency mixer can be used for up and down conversion. In the measurement the first mixer upconverts two quadratures signals used to probe the cavity. In this process the signal with frequency ω_p servers as local oscillator (LO) of the mixer and two intermediate frequencies (IF) or quadrature signals $I(t)$ and $Q(t)$ modulate the signal according to

$$RF(t) = I(t) \cos(\omega_p t) + Q(t) \cos\left(\frac{\pi}{2} + \omega_p t\right) \tag{4.9}$$

in which both IF components have a $\pi/2$ phase shift. The quadratures $I(t)$ and $Q(t)$ are created by a fast arbitrary wave form generator (AWG) with 2.4 Gigasamples

Fig. 4.15 *Left* The quadrature frequency mixer a four port device. *Right* The quadratures I and Q of the signal $A = |A|e^{i\phi}$ with amplitude $|A|$ and phase ϕ

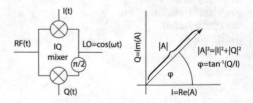

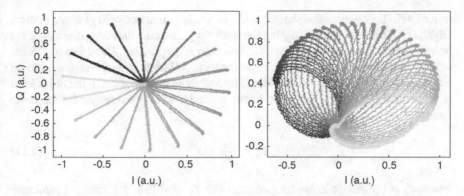

Fig. 4.16 Cavity response in phase space for a short microwave pulse. The signal captures also the cavity ringing during the $10\,\mu s$ long drive pulse. *Left* As the probe tone is in resonace with the cavity ($\omega_p = \omega_c$) the signal is describes a circle when the pase ϕ is varied. *Right* If the cavity probe tone is out of resonance ($\omega_p \neq \omega_c$) this origin of this circle changes

per second (GS/s). In the mixing process the quadratures $I(t)$ and $Q(t)$ are then upconverted by the LO. Therefore the amplitude, pulse duration and phase of the resulting $RF(t)$ signal fed towards the cavity input is controlled.

After the probe tone has passed the cavity, was amplified and has left the cryostat it is down converted again. This is necessary since the real-time measurement of the Gigahertz signal is only possible with very expensive equipment. Therefore the signal is down converted with the reference signal with frequency ω_p used as a local oscillator. The down converted quadratures read then

$$I(t) = RF(t)\cos(\omega_p t) = \frac{1}{2}\left[I(t) + I(t)\cos(2\omega_p t) - Q(t)\sin(2\omega_p t)\right]$$

$$Q(t) = RF(t)\cos\left(\frac{\pi}{2} + \omega_p t\right) = \frac{1}{2}\left[I(t) - I(t)\cos(2\omega_p t) + Q(t)\sin(2\omega_p t)\right]$$

$$(4.10)$$

and after a low pass filter the complex signal $A(t)$ is calculated from both quadratures $I(t)$ and $Q(t)$. The recorded complex signal $A(t) = I(t) + iQ(t)$ of the bare cavity response under the action of a $10\,\mu s$ long rectangular pulse is shown in Fig. 4.16. As the probe tone is in resonance with the cavity ($\omega_p = \omega_c$) the stationary signal describe a circle in phase space. The phase ϕ and amplitude $|A| = \sqrt{I^2 + Q^2}$ of the recorded signal $A = |A|e^{i\phi}$ can be controlled by the first frequency mixer upconverting a pulse waveform send through the cavity.

4.4.2 Signal Averaging

The measured electrical signals are prone to shot and thermal noise. The used cryogenic amplifier is a high electron mobility transistor (HEMT) with a noise temperature of 4 K. This temperature corresponds to a noise signal of ≈20 photons at three Gigahertz. This is extremely low but the +45 dB gain of the first amplifier might not be sufficient to measure weak signals of less then 1000 photons. Therefore additional amplification is added in the measurement chain. Subsequent amplification at room temperature will add additional noise. However in an amplifier chain the noise level is dominated by the first amplifier in the cascade. This is expressed by the noise factor

$$F_n = \frac{T_e}{T_0} + 1 \qquad (4.11)$$

where T_e is the device noise temperature and $T_0 = 290$ K the standard reference temperature. In a cascade of amplifiers the noise factor scales then as

$$F = F_1 + \frac{F_2 - 1}{G_1} + \frac{F_3 - 1}{G_1 G_2} + \frac{F_n - 1}{G_1 G_2 \ldots G_n} + \ldots \qquad (4.12)$$

with each components adding a gain G_n. Therefore it is obvious to put the first amplifier on the 4 K stage to ensure as few as possible noise is added to the measurement signal.

In the presented experiments the typical measurement intensity and mean cavity photon number $|A|^2 = \langle a^\dagger a \rangle$ is in the range of 10^3–10^6 photons. However, a signal corresponding to thousand photons is too small to peak out of the noise floor. In order to overcome this issue the recorded signal is averaged. In Fig. 4.17 the recorded cavity response $|A|^2$ is 10^2–10^6 times averaged. As one can see as the number of averages increases the noise vanishes and the signal to noise ratio increases. The signal to

Fig. 4.17 Time dependent cavity response recorded with 10^2–10^6 averages. The number of averages increases the signal to noise ratio and noise vanishes with averaging

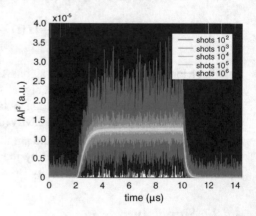

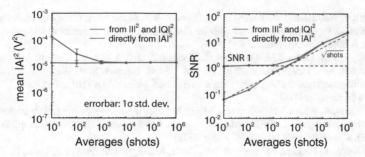

Fig. 4.18 *(Left)* Mean signal and 1σ standard deviation derived with error propagation from $|I|^2$ and $|Q|^2$ or directly from $|A|^2$. *(Right)* The signal to noise ratio scales with the number of averages n as $\sqrt{n}$. The analyzed measurements where taken with intensities corresponding to a mean cavity photon number of 10^3 photons

noise ratio scales with the number of shots n as $\sqrt{n}$. In order to demonstrate this for $|A|^2$ one has to use error propagation

$$\sigma_{|A|^2} = 2\sqrt{|I|^2\sigma_I^2 + |Q|^2\sigma_Q^2} \tag{4.13}$$

with the 1σ standard deviations σ_I and σ_Q of the steady state quadratures signals I and Q. In Fig. 4.18 this behavior is clearly observed. For comparison the signal to noise ratio derived directly from $|A|^2$ is plotted. The deviation from the $\sqrt{n}$ scaling is obvious in the regime where the SNR is smaller than one. From Fig. 4.18 one can see that for a signal of 10^3 photons a number of 10^3 averages is sufficient to achieve signal to noise ratios greater than one. A way to reduce the number of necessary averages is to increase the measurement intensity. Therefore a transmitted signal of 10^6 photons is already resolvable with SNR$\approx$1 in a single measurement shot.

References

1. R. Amsüss, *Strong coupling of an NV spin ensemble to a superconducting resonator*. PhD Thesis, (TU Wien, Austia, 2012)
2. C. Koller. *Towards the experimental realization of hybrid quantum systems*. PhD Thesis, (TU Wien, Austia, 2012)
3. T. Nöbauer. *Sensing, coherent coupling and optimal control with nitrogen-vacancy colour centres in diamond*. PhD Thesis, (TU Wien, Vienna, March 2013)
4. P.K. Day, H.G. LeDuc, B.A. Mazin, A. Vayonakis, J. Zmuidzinas, A broadband superconducting detector suitable for use in large arrays. Nature **425**(6960), 817–821, ISSN 0028-0836 (October 2003)
5. A. Wallraff, D.I. Schuster, A. Blais, L. Frunzio, R.-S. Huang, J. Majer, S. Kumar, S.M. Girvin, R.J. Schoelkopf, Strong coupling of a single photon to a superconducting qubit using circuit quantum electrodynamics. Nature **431**, 162–167 (2004)

6. A. Blais, R.S. Huang, A. Wallraff, S.M. Girvin, R.J. Schoelkopf, Cavity quantum electrody-namics for superconducting electrical circuits: an architecture for quantum computation. Phys. Rev. A **69**(6), 062320 (June 2004)
7. J. Majer, J.M. Chow, J.M. Gambetta, J. Koch, B.R. Johnson, J.A. Schreier, L. Frunzio, D.I. Schuster, A.A. Houck, A. Wallraff, A. Blais, M.H. Devoret, S.M. Girvin, R.J. Schoelkopf, Coupling superconducting qubits via a cavity bus. Nature **449**(7161), 443–447, ISSN 0028-0836 (September 2007)
8. L. Frunzio, A. Wallraff, D. Schuster, J. Majer, R. Schoelkopf, Fabrication and characterization of superconducting circuit QED devices for quantum computation. IEEE Trans. Appl. Supercond. **15**(2), 860–863, ISSN 1051-8223 (June 2005)
9. M. Göppl, A. Fragner, M. Baur, R. Bianchetti, S. Filipp, J.M. Fink, P.J. Leek, G. Puebla, L. Steffen, A. Wallraff, Coplanar waveguide resonators for circuit quantum electrodynamics. J. Appl. Phys. **104**(11), 113904, ISSN 0021-8979, 1089-7550 (December 2008)
10. R. Simons. Suspended slot line using double layer dielectric. Trans. Microw. Theory Tech. IEEE, **29**(10) (1981)
11. Y. Kubo, F.R. Ong, P. Bertet, D. Vion, V. Jacques, D. Zheng, A. Dréau, J.-F. Roch, A. Auffeves, F. Jelezko, J. Wrachtrup, M.F. Barthe, P. Bergonzo, D. Esteve, Strong coupling of a spin ensemble to a superconducting resonator. Phys. Rev. Lett. **105**(14), 140502 (2010)
12. R. Amsüss, Ch. Koller, T. Nöbauer, S. Putz, S. Rotter, K. Sandner, S. Schneider, M. Schramböck, G. Steinhauser, H. Ritsch, J. Schmiedmayer, J. Majer, Cavity QED with magnetically coupled collective spin states. Phys. Rev. Lett. **107**(6), 060502 (2011)
13. X. Zhu, S. Saito, A. Kemp, K. Kakuyanagi, S. Karimoto, H. Nakano, W.J. Munro, Y. Tokura, M. Everitt, K.S. Nemoto, M. Kasu, N. Mizuochi, K. Semba. Coherent coupling of a supercon-ducting flux qubit to an electron spin ensemble in diamond. Nature **478**, 221–224 (2011)
14. D.I. Schuster, A.P. Sears, E. Ginossar, L. DiCarlo, L. Frunzio, J.J.L. Morton, H. Wu, G.A.D. Briggs, B.B. Buckley, D.D. Awschalom, R.J. Schoelkopf, High-cooperativity coupling of electron-spin ensembles to superconducting cavities. Phys. Rev. Lett. **105**(14), 140501 (2010)
15. S. Probst, H. Rotzinger, S. Wünsch, P. Jung, M. Jerger, M. Siegel, A.V. Ustinov, P.A. Bushev, Anisotropic rare-earth spin ensemble strongly coupled to a superconducting resonator. Phys. Rev. Lett. **110**(15), 157001 (2013)
16. C.W. Zollitsch, K. Mueller, D.P. Franke, S.T.B. Goennenwein, M.S. Brandt, R. Gross, H. Huebl, High cooperativity coupling between a phosphorus donor spin ensemble and a superconducting microwave resonator. Appl. Phys. Lett. **107**(14), 142105, ISSN 0003-6951, 1077-3118 (October 2015)
17. D.A. Redman, S. Brown, R.H. Sands, S.C. Rand, Spin dynamics and electronic states of N-V centers in diamond by EPR and four-wave-mixing spectroscopy. Phys. Rev. Lett. **67**(24), 3420–3423 (1991)
18. F. Jelezko, T. Gaebel, I. Popa, A. Gruber, J. Wrachtrup, Observation of coherent oscillations in a single electron spin. Phys. Rev. Lett. **92**(7), 076401 (2004)
19. F. Jelezko, J. Wrachtrup, Single defect centres in diamond: a review. Phys. Status Solidi A Appl. Mater. Sci. **203**(13), 3207 (2006)
20. L. Childress, M.V. Gurudev Dutt, J.M. Taylor, A.S. Zibrov, F. Jelezko, J. Wrachtrup, P.R. Hemmer, M.D. Lukin, Coherent dynamics of coupled electron and nuclear spin qubits in diamond. Science **314**(5797), 281–285 (2006)
21. Y. Kubo, C. Grèzes, A. Dewes, T. Umeda, J. Isoya, H. Sumiya, N. Morishita, H. Abe, S. Onoda, T. Ohshima, V. Jacques, A. Dréau, J.-F. Roch, I. Diniz, A. Auffeves, D. Vion, D. Esteve, P. Bertet, Hybrid quantum circuit with a superconducting qubit coupled to a spin ensemble. Phys. Rev. Lett. **107**(22), 220501 (2011)
22. N.V. Prokof'ev, P.C.E. Stamp. Theory of the spin bath. Rep. Prog. Phys. **63**(4), 669, ISSN 0034-4885 (2000)
23. P.L. Stanwix, L.M. Pham, J.R. Maze, D. Le Sage, T.K. Yeung, P. Cappellaro, P.R. Hemmer, A. Yacoby, M.D. Lukin, R.L. Walsworth, Coherence of nitrogen-vacancy electronic spin ensem-bles in diamond. Phys. Rev. B **82**(20), 201201 (2010)

24. A. Jarmola, V.M. Acosta, K. Jensen, S. Chemerisov, D. Budker, Temperature- and magnetic-field-dependent longitudinal spin relaxation in nitrogen-vacancy ensembles in diamond. Phys. Rev. Lett. **108**(19), 197601 (2012)
25. N. B.-Gill, L.M. Pham, A. Jarmola, D. Budker, R.L. Walsworth. Solid-state electronic spin coherence time approaching one second. Nat. Commun. **4**(1743), (April 2013)
26. V.M. Acosta, E. Bauch, M.P. Ledbetter, C. Santori, K.-M.C. Fu, P.E. Barclay, R.G. Beausoleil, H. Linget, J.F. Roch, F. Treussart, S. Chemerisov, W. Gawlik, D. Budker, Diamonds with a high density of nitrogen-vacancy centers for magnetometry applications. Phys. Rev. B **80**(11), 115202 (2009)
27. T. Nöbauer, K. Buczak, A. Angerer, S. Putz, G. Steinhauser, J. Akbarzadeh, H. Peterlik, J. Majer, J. Schmiedmayer, M. Trupke, Creation of ensembles of nitrogen-vacancy centers in diamond by neutron and electron irradiation. arXiv:1309.0453, (2013)

Chapter 5
Collective Spin States Coupled to a Single Mode Cavity—Strong Coupling

In the theoretical part of this thesis I described in great detail the electromagnetic field of single mode cavities and the interaction with the magnetic moment of single or ensembles of electron spins. In this chapter I will show the experimental realization of strong coupling [1, 2] in a solid-state cavity QED system consisting of a superconducting microwave cavity and an ensemble of electron spins hosted by nitrogen vacancy (NV) spins. The main result of strong coupling described in this chapter has been demonstrated in earlier work [3, 4] and lays the foundation of the performed measurements carried out in this thesis. Strong coupling to an ensemble of electron spins in a solid state crystal has also been demonstrate by many other research groups recently, also with NV spins [5, 6] and various other spin systems [7–10].

5.1 Strong Coupling

The strong interaction in the experiment relies on the basic theoretical framework of the Tavis-Cummings [11] and the Dicke [12] model, in which the cavity spin interaction is collectively enhanced. The typical cavity and single spin coupling strength (i.e. vacuum Rabi frequency) in the presented experiment of $\lesssim 10\,\mathrm{Hz}$ [13, 14] is rather small, therefore a large number $N = 10^{12}$ of weakly dipole interacting spins in the cavity mode volume is needed to advance into the strong coupling regime of cavity QED. Such an ensemble of "non" interacting two-level systems coupled to a single mode cavity is described by a standard Tavis-Cummings [11] Hamiltonian which reads in the rotating wave approximation

$$\mathcal{H} = \hbar\omega_c a^\dagger a + \frac{\hbar}{2}\sum_{j=1}^{N}\omega_j\sigma_j^z + \hbar\sum_{j=1}^{N}g_j\left[\sigma_j^- a^\dagger + \sigma_j^+ a\right] + \mathrm{i}\eta(a^\dagger\mathrm{e}^{-\mathrm{i}\omega_p t} - a\mathrm{e}^{\mathrm{i}\omega_p t}), \quad (5.1)$$

© Springer International Publishing AG 2017
S. Putz, *Circuit Cavity QED with Macroscopic Solid-State Spin Ensembles*,
Springer Theses, DOI 10.1007/978-3-319-66447-7_5

with bosonic creation (annihilation) operators $a^\dagger$ (a) standing for the cavity mode with frequency ω_c. The Pauli spin operators $\sigma_j^{\pm,z}$ are associated with the jth spin with frequency ω_j in a macroscopic spin ensemble of N spins. This Hamiltonian includes a driving term with amplitude η and frequency ω_p which allows to stimulate and measure the system response. The inhomogeneous spectral spin broadening results in a variation of the spin transition frequencies ω_j centered around a central spin frequency ω_s. It is convenient and useful to introduce a frequency de-tuning parameter for the central spin frequency (i.e. the ensemble) and the cavity resonance frequency $\Delta = \omega_s - \omega_c$.

5.1.1 "Vacuum" Rabi Splitting

In the presented experiments the central spin frequency ω_s of the spin ensemble is Zeeman shifted into resonance with the cavity ($\omega_s = \omega_c$) by applying an external d.c. magnetic field. On resonance ($\Delta = 0$) the collectively enhanced interaction $\Omega_R/2 \approx \Omega = \sqrt{\sum_j^N g_j^2}$ scales approximately with the number of spins as $\sqrt{N}$. Application of the collective spin operators $J^\pm$ on the spin ground state $|G\rangle$, creates a symmetric spin state $|B\rangle = J^+|G\rangle$ which gives rise to two polariton modes

$$|\pm\rangle = \frac{1}{\sqrt{2}}(|1, G\rangle \pm |0, B\rangle) \tag{5.2}$$

which are a superstition and maximally entangled state between the cavity mode and a collective spin state, as is discussed in Chap. 3. The polariton mode includes a spin state in the single excitation manifold of the Dicke model which is justified only when the bosonisation or Holstein-Primakoff approximation [15] is applicable. One should note that the expression of the polariton modes follow the dressed state picture and spin and cavity states are denoted as number or Fock states. In the experiments carried out in the latter the system is probed with weak coherent microwave signals, but since only expectation values of the cavity field are measured in transmission we assume the cavity amplitude $|A|^2 = \langle a^\dagger a \rangle = N_{\text{Photons}}$.

The hybridization of the cavity spin system is probed by performing continuous wave transmission spectroscopy measurements on the cavity with a vector network analyzer (VNA). The external d.c. magnetic field $\vec{B} = (|B|\sin\phi, |B|\cos\phi, 0)$ scanned by varying the amplitude $|B|$ and field direction ϕ, which allows to Zeeman shift spins in and out of resonance with the cavity. The magnetic field alignment with respect to the NV axis controls the tuning of the individual NV subensembles. It is possible to tune two and all four NV subensembles in resonance with the cavity at once. In Fig. 5.1 the magnetic field angle is set to $\phi = 45°$ and the magnetic field amplitude $|B|$ is scanned. At zero external magnetic field $|B| = 0$ the bare cavity resonance is observed with a linewidth $\kappa/2\pi = 440 \pm 5\,\text{kHz}$ (HWHM) and resonance frequency $\omega_c/2\pi = 2.691\,\text{GHz}$. As the magnetic field strength increases

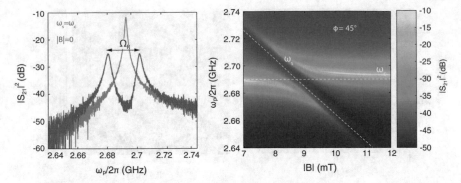

Fig. 5.1 At zero external field a bare cavity resonance with frequency $\omega_c/2\pi = 2.691$ GHz and $\kappa/2\pi = 440 \pm 5$ kHz (HWHM) is observed. When the spin ensemble is Zeeman shifted into resonance with the cavity a collectively enhanced coupling strength $\Omega_R/2\pi = 19 \pm 0.01$ MHz with a total decoherence rate $\Gamma/2\pi = 3.1$ MHz (FWHM) is measured. The preferred magnetic field alignment in the presented experiments is $\phi = 45°$, since at this angle the NV ensemble separation is largest while the magnetic filed is maximally aligned with the NV quantization axis

two degenerate NV subensembles are tuned into resonance with the cavity and at $\omega_c = \omega_s$ a clear Rabi splitting $\Omega_R/\pi = 19 \pm 0.01$ MHz is observed with a cooperativity of $C \approx 18$. The Rabi splitting is a direct measure of the collectively enhanced interaction strength since $\Omega_R \approx \Omega/2$. The peaks in the transmission spectra shown in Fig. 5.1 correspond to the polariton modes of the coupled system and have a full-width at half-maximum of $\Gamma/2\pi = 3.1$ MHz. For an explanation of cooperativity C in such a cavity QED system see Chap. 8. This high cooperativity and large collectively enhanced coupling strength fulfills the criteria: $\Omega \gg \Gamma + \kappa$ and allows to operate the system in the strong coupling regime.

As described the collective enhanced coupling strength $\Omega_R \approx \Omega/2$ scales with the number of spins N as $\sqrt{N}$. This behavior can be tested by changing the magnetic field angle from $\phi = 45°$ to $\phi = 0°$ at which two and all four NV subensembles can be brought into resonance with the cavity, respectively. In Fig. 5.2 the Zeeman tuning for two different magnetic field angles $\phi = 0°$ and $\phi = 3.5°$ is shown, for the latter the NV subensembles become non degenerate as is expected. The Rabi splittings in Fig. 5.3 are measured for two magnetic field angles $\phi = 45°$ and $\phi = 0°$ at which the NV subensembles are well separated and degenerate, respectively. The Rabi splitting $\Omega_R/\pi = 19$ MHz observed at $\phi = 45°$ is enhanced to $\Omega_R/\pi = 26.6$ MHz at $\phi = 0°$, which corresponds approximately to a $\sqrt{2}$ enhancement and demonstrates the $\sqrt{N}$ scaling in the experiment.

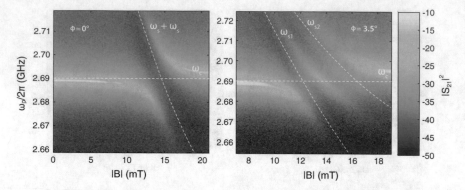

Fig. 5.2 Cavity transmission measurements for two scans of the magnetic field amplitude $|B|$ for two different magnetic field angles $\phi = 0°$ and $\phi = 3.5°$. *Left* all four NV subensembles are degenerate and couple as one spin ensemble to the cavity. *Right* as the magnetic field angle is slightly tilted NV spin subensembles become non-degenerate which clearly becomes visible in the spectroscopic measurements

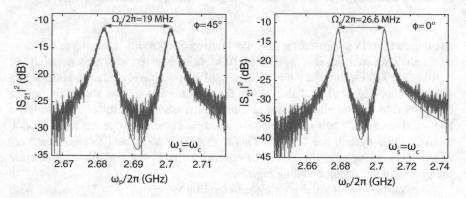

Fig. 5.3 The collectively enhanced coupling Ω scales as $\sqrt{N}$. The magnetic field angle *(left)* $\phi = 45°$ and *(right)* $\phi = 0°$ shift two and all four subensembles into resonance. The enhanced Rabi splitting demonstrates the $\sqrt{N}$ scaling

5.1.2   Dispersive Measurements

If the cavity and central spin detuning $\Delta = \omega_s - \omega_c$ becomes large compared to collective coupling strength $\Delta \gg \Omega$ there is no direct energy exchange between cavity and spin ensemble. Nevertheless, due to the strong interaction the bare and off resonant cavity resonance with frequency ω_c is shifted by

$$\Delta \approx \pm \frac{\Omega^2}{\Delta} \tag{5.3}$$

in the dispersive regime. This cavity shift can be used to estimate some important quantities: the spectral spin linewidth γ_{inh}, shape $\rho(\omega)$ and spin relaxation time T_1. A simple scheme to measure these quantities is to inject a strong off resonant driving field ($\omega_p \neq \omega_c$) first, and to monitor the cavity resonance frequency by transmission spectroscopic measurements afterwards. If the strong probe tone excites the de-tuned spin ensemble (i.e. $\omega_p = \omega_s$) the ensemble saturates and the dispersive shift is canceled. This principle allows to reconstruct the spin ensemble line shape and width by scanning the probe tone frequency on a sufficient slow time scale in order to ensure spin relaxation between each scanned frequency point. In the left panel of Fig. 5.4 such a measurement is shown, the NV spin ensemble is $\Delta/2\pi \approx 150$ MHz de-tuned which is in good agreement with the maximal observed dispersive cavity shift of $\frac{\Omega^2}{2\pi\Delta} \approx 650$ kHz. As the probe tone frequency is scanned the linewidth γ_{inh} can be estimated from the dispersive cavity shift. The errorbars correspond to the half cavity linewidth κ measured by transmission spectroscopy measurements and estimated by a Lorentzian fit at each scanned frequencies point to determine the shifted cavity resonance frequency.

With this method the inhomogeneous spectral line shape $\rho(\omega)$ can be reconstructed with great accuracy and the spin ensemble in the presented experiment is best described by a q-Gaussian function [16]

$$\rho(\omega) = \left[1 - (1 - q) \frac{(\omega - \omega_s)^2}{\Delta^2} \right]^{\frac{1}{1-q}}, \tag{5.4}$$

characterized by the dimensionless shape parameter $q = 1.39$. This function yields the convolution of a Lorentzian and Gaussian distribution which is controlled by the parameter q. The full-width at half-maximum of $\rho(\omega)$ is given by $\gamma_{inh} = 2\Delta\sqrt{\dfrac{2^q - 2}{2q - 2}}$ with a line-width $\gamma_{inh}/2\pi = 9.4$ MHz (FWHM) [16, 17].

Furthermore the spin ensemble relaxation time is estimated by probing the spin ensemble at the central spin frequency ω_s. At this frequency the dispersive shift is maximal and ensures maximal signal to noise ratios. The dispersive cavity shift is monitored over time and as the saturated spin ensemble relaxes back into its ground state the dispersive cavity shift exponentially builds up again. In the right panel of Fig. 5.4 an exponential fit yields a relaxation time constant $T_1 = 170.2 \pm 13.5$ s. However it must be stressed that this is not the true T_1 time of the NV spin ensemble. This follows from the simple fact that due to the non uniform single spin Rabi frequencies there is a clear direction for spin diffusion within the ensemble which limits and the shortens the spin life time T_1.

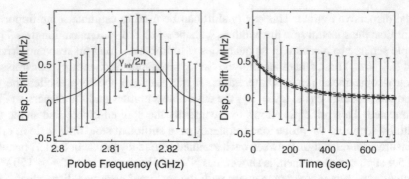

Fig. 5.4 *Left* Dispersive shift of the cavity resonance frequency when the NV spin ensemble is $\Delta/2\pi \approx 150\,$MHz de-tuned from the resonator. The observed peak corresponds the transition of the inhomogeneously broadened spin ensemble. *Right* The time dependent cavity shift can also be uses to estimate the spin relaxation time $T_1 = 170.2 \pm 13.5\,$s

5.2 Rabi Oscillations

In the previous section the Rabi splitting was observed by spectroscopic steady state measurements. This hybridization allows the coherent exchange of energy between the cavity and the spin ensemble via the polariton modes, a phenomena known as Rabi oscillations [18–20]. To show such oscillations experimentally, measurements have to be performed in the time domain. This is possible by homodyne quadrature measurements in which a high frequency signal transmitted through the cavity is down converted and analyzed as discussed in Sect. 4.4.1. In the latter the Zeeman shift of the spin ensemble is always adjusted in such a way that only tow NV subnesemble are in resonance with the cavity hence $\omega_p = \omega_c$, which corresponds to a magnetic field alignment of $\phi = 45°$.

5.2.1 Linear Rabi Oscillations

The dynamical response of the system is probed by scanning the drive frequency ω_p of a rectangular microwave pulse and measuring the time dependent transmitted intensity $|A(\omega_p, t)|^2$ through the cavity. If a rectangular drive signal with amplitude $\eta(t)$, sufficiently long to drive the system into its steady state, is applied and the probe frequency is scanned the time resolved Rabi splitting is observed in the time dependent cavity transmission as is shown in the left panel of Fig. 5.5. If the probe tone becomes resonant with the cavity and the spin ensemble ($\omega_c = \omega_s = \omega_p$) the signal will exhibit cohrente oscillations, which settle towards a stationary state. As is shown in the right panel of Fig. 5.5 after the drive is switched off, coherent oscillations with frequency Ω_R are observed. These oscillation can be considered as an interference of the polariton modes $|\pm\rangle$, oscillating clock and counter clockwise with respect to the probe tone in phase space.

This coherent oscillation are known as Rabi oscillation in which an excitation oscillates back and forth between cavity and spin ensemble. Therefore it is most instructive to make some simple considerations on how the coupled system evolves in time in the dressed state picture

$$|\pm\rangle = \frac{1}{\sqrt{2}}(|1, G\rangle \pm |0, B\rangle) \tag{5.5}$$

by making a basic time evolution of the polariton modes. Both sates are separated in energy and degenerate due to the Rabi splitting by

$$\Delta\mathcal{E} = \hbar\Omega_R \tag{5.6}$$

assuming cavity and spins on resonance and $\Delta = \omega_c - \omega_s = 0$. A linear combination of both dressed states

$$|1, G\rangle \quad = |+\rangle + |-\rangle \text{ and } |0, B\rangle \quad = |+\rangle - |-\rangle \tag{5.7}$$

give states for which a single excitation is entirely in the spin ensemble or cavity mode. A time evolution of the dressed states shows then how an excitation in the hybridized system evolves

$$|1, G\rangle = e^{-i\frac{\Omega_R}{2}t}|+\rangle + e^{i\frac{\Omega_R}{2}t}|-\rangle \tag{5.8}$$

$$|0, B\rangle = e^{-i\frac{\Omega_R}{2}t}|+\rangle - e^{i\frac{\Omega_R}{2}t}|-\rangle \tag{5.9}$$

back and forth between the cavity and the spin ensemble. After each integer multiple n and time $t = \frac{2\pi(2n+1)}{\Omega_R}$ the system in an initial state $|1, G\rangle$ has evolved into a state $|0, B\rangle$ and an excitation is fully transferred from the cavity mode to the spin ensemble and vice versa.

In an input output formalism the time evolution of the coupled cavity spin system is derived from the Heisenberg operator equations

$$\dot{a} = \frac{i}{\hbar}[\mathcal{H}, a] - \kappa a \text{ and } \dot{\sigma_j^-} = \frac{i}{\hbar}[\mathcal{H}, \sigma_j^-] - \gamma_\perp \sigma_j^- \tag{5.10}$$

with cavity κ and spin γ dissipation rates and by a semi classical approach as discussed later on [21–23]. The expectation value of the cavity operator is then mapped to its experimental counterpart the transmitted intensity through the cavity since $|\langle a^\dagger a\rangle|^2 = |A|^2$. Oscillations observed in the transmitted cavity intensity can be modeled by this approach and in the linear regime can be treated as a collection of coupled harmonic oscillators [17, 24]. In the experiment the system is probed with low intensities of typically $|A|^2 > 1000$ photons or 10^{-9} photons per spin in the cavity on average.

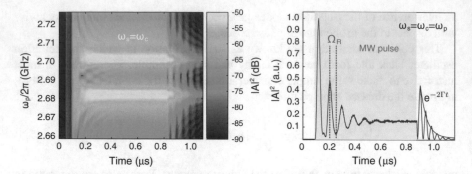

Fig. 5.5 Time resolved Rabi splitting *(left)* by scanning the probe frequency ω_p and $\omega_p = \omega_c$. If the cavity probe field is in resonance with the hybridize system ($\omega_p = \omega_c = \omega_s$) Rabi oscillation are observed in the transmitted intensity. The system starts to oscillate and and decays then towards a stationary state. After the drive is switched off Rabi oscillations occur again with frequency Ω_R and decay with a time constant Γ. The observed overshoot of the first two Rabi peaks is due an energy stored in the spin ensemble which is released back into the cavity and interferes with the cavity field

Such a weak driving ensures that the Holstein-Primakoff or bosonic approximation is valid and the derived linear equations of motion for spins and cavity amplitude are accurate and can be interpret as coupled harmonic oscillators. A toy model of two coupled harmonic oscillators can be used from which a solution can be found by treating the degeneracy of the polartion modes as a normal mode splitting. This regime is called the linear regime in the experiment, but one has to note that the oscillation frequency Ω_R is constant. Compared to the semi-classical Rabi model the Rabi frequency does not scale with the drive intensity in this linear regime. The Rabi oscillations will decay with a time constant Γ corresponding to the width of the polariton modes and is determined by the cavity κ and the spin γ dissipation rates as well as the inhomogeneous spin broadening. How the total decoherence scales in such a system when inhomogeneous spin broadening is present will be discussed in the following chapter.

5.2.2   Non-linear Rabi Oscillations

In the previous section the carried out measurements used weak probe intensities (10^3–10^6 photons on average in the cavity) to ensure that the number of photons in the cavity is well below the number of spins $N \approx 10^{12}$ in the cavity mode volume. If the power levels are this weak the system is in the linear regime and behaves similar to two or collection of coupled harmonic oscillators. As the drive power and amplitude η is increased the system leave this regime and non-linear effects can be observed directly in the transmitted intensity through the cavity which is observable in the Rabi oscillations. This can be understand as the spin ensemble is excited with stronger

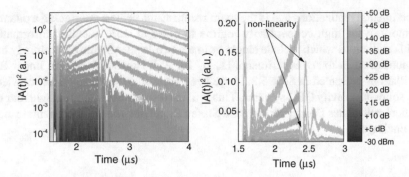

Fig. 5.6 *Left* The stimulated system shows Rabi oscillation and is further bleached as the drive intensity $\propto \eta^2$ is scanned over a range of 47 dB. In the intermediate power regime *right* where the spin ensemble starts to bleach non-linear Rabi oscillations are observed as the Rabi peaks have non-monotonic peak amplitudes

and stronger driving fields the Dicke ladder is subsequently climbed. This means the one excitation manifold of the Dicke model is left and the collective enhanced coupling strength Ω increase, which also means that the oscillation frequency Ω_R changes while the system undergoes Rabi oscillations.

In Fig. 5.6 the probe pulse intensity is scanned from -30 dBm ($\approx 10^6$ photons) to $+17$ dBm ($\approx 10^{11}$ photons) on average per pulse in the cavity. The single shot measurements presented in Fig. 5.6 show the transmitted intensity $|A|^2$ through the cavity. For low pulse intensities the system is in the linear regime but as the drive power is increased the Rabi oscillations show first non-linear effects, which manifests in the non-monotonic Rabi peak amplitudes. As the drive power is increased further, the transmission amplitude rises but also the Rabi oscillations completely vanish since the spin ensemble is saturated and bleached. This means instead of climbing the Dicke ladder the spin ensemble is driven into the spin dark or subradiant subspace. This prevents effects such as super absorbance and radiance and consequently coherently inverting the spin ensemble in the presented experiment. This is attributed to the effect of the inhomogeneous spectral spin broadening and inhomogeneous single spin-cavity coupling strengths. Therefore making it impossible to excite all spins in phase and to preserve the symmetric Dicke spin state.

5.3 Conclusion

The presented measurements and data sets demonstrate the ground lying principles of cavity QED which are strong coupling and Rabi oscillations. The strong light matter interaction gives rise to the formation of polariton modes and allows to coherently exchange energy between spins ensemble and the cavity with the Rabi frequency Ω_R. This hybridization and strong coupling has been discussed in many contexts and experimental realizations and is the main mechanisms behind interfacing a quantum memory in the high cooperativity regime in order to store quantum information. Rabi oscillations which include single or many two-level systems have already been demonstrated in many experiments [1, 6, 18, 25]. However, the non linear Rabi oscillations presented in Sect. 5.2.2 have been unresolved yet to my best knowledge in a solid-state cavity QED systems. This is a first result in collaboration with our theoretical colleagues Dmitry Krimer and Stefan Rotter and more experimental studies are envisaged.

References

1. G. Rempe, R.J. Thompson, R.J. Brecha, W.D. Lee, H.J. Kimble, Optical bistability and photon statistics in cavity quantum electrodynamics. Phys. Rev. Lett. **67**(13), 1727–1730 (1991)
2. R.J. Thompson, G. Rempe, H.J. Kimble, Observation of normal-mode splitting for an atom in an optical cavity. Phys. Rev. Lett. **68**(8), 1132–1135 (1992)
3. R. Amsüss, Ch. Koller, T. Nöbauer, S. Putz, S. Rotter, K. Sandner, S. Schneider, M. Schramböck, G. Steinhauser, H. Ritsch, J. Schmiedmayer, J. Majer, Cavity QED with magnetically coupled collective spin states. Phys. Rev. Lett. **107**(6), 060502 (2011)
4. R. Amsüss. *Strong coupling of an NV spin ensemble to a superconducting resonator.* PhD Thesis, (TU Wien, Austia, 2012)
5. Y. Kubo, F.R. Ong, P. Bertet, D. Vion, V. Jacques, D. Zheng, A. Dréau, J.-F. Roch, A. Auffeves, F. Jelezko, J. Wrachtrup, M.F. Barthe, P. Bergonzo, D. Esteve, Strong coupling of a spin ensemble to a superconducting resonator. Phys. Rev. Lett. **105**(14), 140502 (2010)
6. X. Zhu, S. Saito, A. Kemp, K. Kakuyanagi, S. Karimoto, H. Nakano, W.J. Munro, Y. Tokura, M. Everitt, K.S. Nemoto, M. Kasu, N. Mizuochi, K. Semba, Coherent coupling of a superconducting flux qubit to an electron spin ensemble in diamond. Nature **478**, 221–224 (2011)
7. D.I. Schuster, A.P. Sears, E. Ginossar, L. DiCarlo, L. Frunzio, J.J.L. Morton, H. Wu, G.A.D. Briggs, B.B. Buckley, D.D. Awschalom, R.J. Schoelkopf, High-cooperativity coupling of electron-spin ensembles to superconducting cavities. Phys. Rev. Lett. **105**(14), 140501 (2010)
8. S. Probst, H. Rotzinger, S. Wünsch, P. Jung, M. Jerger, M. Siegel, A.V. Ustinov, P.A. Bushev, Anisotropic rare-earth spin ensemble strongly coupled to a superconducting resonator. Phys. Rev. Lett. **110**(15), 157001 (2013)
9. C.W. Zollitsch, K. Mueller, D.P. Franke, S.T.B. Goennenwein, M.S. Brandt, R. Gross, H. Huebl, High cooperativity coupling between a phosphorus donor spin ensemble and a superconducting microwave resonator. Appl. Phys. Lett. **107**(14):142105, (October 2015). ISSN 0003-6951, 1077-3118
10. Y. Tabuchi, S. Ishino, A. Noguchi, T. Ishikawa, R. Yamazaki, K. Usami, Y. Nakamura, Coherent coupling between a ferromagnetic magnon and a superconducting qubit. Science, **349**(6246):405–408 (July 2015). ISSN 0036-8075, 1095-9203
11. M. Tavis, F.W. Cummings, Exact solution for an n-molecule–radiation-field hamiltonian. Phys. Rev. **170**(2), 379–384 (1968)
12. R.H. Dicke, Coherence in spontaneous radiation processes. Phys. Rev. **93**(1), 99–110 (1954)
13. J. Verdú, H. Zoubi, Ch. Koller, J. Majer, H. Ritsch, J. Schmiedmayer, Strong magnetic coupling of an ultracold gas to a superconducting waveguide cavity. Phys. Rev. Lett. **103**(4), 043603 (2009)
14. A. İmamoğlu, Cavity QED based on collective magnetic dipole coupling: spin ensembles as hybrid two-level systems. Phys. Rev. Lett. **102**(8), 083602 (2009)
15. H. Primakoff, T. Holstein, Many-body interactions in atomic and nuclear systems. Phys. Rev. **55**(12), 1218–1234 (1939)

16. K. Sandner, H. Ritsch, R. Amsüss, Ch. Koller, T. Nöbauer, S. Putz, J. Schmiedmayer, J. Majer, Strong magnetic coupling of an inhomogeneous nitrogen-vacancy ensemble to a cavity. Phys. Rev. A **85**(5), 053806 (2012)

17. D.O. Krimer, S. Putz, J. Majer, S. Rotter, Non-Markovian dynamics of a single-mode cavity strongly coupled to an inhomogeneously broadened spin ensemble. Phys. Rev. A **90**(4), 043852 (2014)

18. M. Brune, F. Schmidt-Kaler, A. Maali, J. Dreyer, E. Hagley, J.M. Raimond, S. Haroche, Quantum rabi oscillation: a direct test of field quantization in a cavity. Phys. Rev. Lett. **76**(11), 1800–1803 (1996)

19. T.B. Norris, J.-K. Rhee, C.-Y. Sung, Y. Arakawa, M. Nishioka, C. Weisbuch, Time-resolved vacuum Rabi oscillations in a semiconductor quantum microcavity. Phys. Rev. B **50**(19), 14663–14666 (1994)

20. T. Yoshie, A. Scherer, J. Hendrickson, G. Khitrova, H.M. Gibbs, G. Rupper, C. Ell, O.B. Shchekin, D.G. Deppe, Vacuum Rabi splitting with a single quantum dot in a photonic crystal nanocavity. Nature, **432**(7014):200–203, (November 2004). ISSN 0028-0836

21. J. Gripp, S.L. Mielke, L.A. Orozco, Evolution of the vacuum Rabi peaks in a detuned atom-cavity system. Phys. Rev. A **56**(4), 3262–3273 (1997)

22. P.D. Drummond, Optical bistability in a radially varying mode. IEEE J. Quantum Electron. **17**(3):301–306, (March 1981). ISSN 0018-9197

23. M.J. Martin, D. Meiser, J.W. Thomsen, J. Ye, M.J. Holland, Extreme nonlinear response of ultranarrow optical transitions in cavity QED for laser stabilization. Phys. Rev. A, **84**(6):063813, (December 2011)

24. I. Diniz, S. Portolan, R. Ferreira, J.M. Gérard, P. Bertet, A. Auffèves, Strongly coupling a cavity to inhomogeneous ensembles of emitters: potential for long-lived solid-state quantum memories. Phys. Rev. A **84**(6), 063810 (2011)

25. Y. Kubo, I. Diniz, A. Dewes, V. Jacques, A. Dréau, J.-F. Roch, A. Auffeves, D. Vion, D. Esteve, P. Bertet, Storage and retrieval of a microwave field in a spin ensemble. Phys. Rev. A **85**(1), 012333 (2012)

Chapter 6
Spin Ensembles and Decoherence in the Strong-Coupling Regime—Cavity Protection

This chapter is based on the following two research papers:

- **"Protecting a spin ensemble against decoherence in the strong-coupling regime of cavity QED"** S. Putz, D. O. Krimer, R. Amsüss, A. Valookaran, T. Nöbauer, J. Schmiedmayer, S. Rotter and J. Majer
  Nature Physics 10, 720–724 (2014)–Published online 17 August 2014
- **"Non-Markovian dynamics of a single-mode cavity strongly coupled to an inhomogeneously broadened spin ensemble"** D. O. Krimer, S. Putz, J. Majer, and S. Rotter
  Phys. Rev. A 90, 043852–Published 24 October 2014

and uses parts of it. These two publications were a joint collaboration with my fellow colleagues from the theory department of the TU Wien and draw from their theoretical expertise.[1] The presented results laid the conceptual ground work for many experiments carried out in this thesis.

6.1 Introduction

As shown in the previous chapter the strong coupling regime of cavity QED is in reach by collective enhanced cavity spin interactions [1–5]. However this enhancement comes with a considerable downside, namely inhomogeneous line broadening of the spin ensemble [6–9]. In these systems spectral broadening limits the coherence times of the system and the performance of the coherent transfer and storage of quantum information [10, 11]. In particular for an ensemble of electron spins magnetic dipolar interactions with excess nuclear and electron spins in the host crystal and additional lattice stress act as the dominant source of spin dephasing. Active techniques such as spin echos [12, 13] have been used to overcome this limitation, but require a

---

[1] I do not claim to have performed the numerical simulations capturing the dynamical system response presented in the latter.

© Springer International Publishing AG 2017
S. Putz, *Circuit Cavity QED with Macroscopic Solid-State Spin Ensembles*,
Springer Theses, DOI 10.1007/978-3-319-66447-7_6

large experimental overhead. An alternative way to circumvent this limitation has been pointed out in theoretical proposals [6, 7] recently. If the collective enhanced coupling strength is large the specific line shape of the inhomogeneous spectral spin distribution acts beneficiary on the total system decay rate. As a matter of fact this can be used to suppress the influence of inhomogeneous spin broadening without active spin manipulation. In the latter I will briefly discuss the principle of this phenomena and show the experimental verification of this effect called the "cavity protection effect" [6, 7, 9, 14].

6.2 The Principle of Cavity Protection

As was shown in Chap. 3 the collective enhanced coupling strength Ω scales with the number of spins N in the cavity as $\sqrt{N}$. If we treat the ensemble as non-interacting individual spins, we can assume that the inhomogeneous linewidth γ_{inh} will not change if the number of total spins is increased and only the spectral density of spins or subradiant states rises. Therefore we immediately would argue that the line width of the polariton modes Γ should remain constant and only the Rabi splitting Ω_R increases with N. In Chap. 5 the $\sqrt{N}$ scaling of Ω_R was demonstrated. The enhancement of the number of spins by a factor of two, by tuning two and then all four NV subensembles into resonance with the cavity, shows the $\sqrt{N}$ scaling. This measurement is shown again in Fig. 6.1 where the Rabi splitting scales as $\sqrt{2}$ as expected. However a surprising effect is immediately visible, the line width of the polaritonic peaks Γ is reduced as the Rabi splitting is increased.

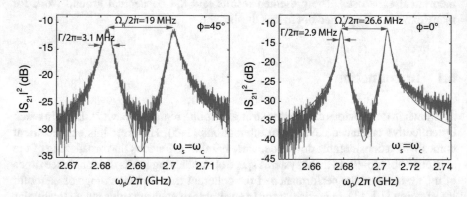

Fig. 6.1 *Left* Two of the four NV sub ensembles are tuned into resonance with the cavity by applying an external Zeeman field aligned and tilted by 45° in the (100) crystal plane. *Right* All four NV sub ensembles are tuned into resonance with the cavity by applying an external Zeeman field aligned and tilted by 0° in the (100) crystal plane. The vacuum Rabi splitting is increased by $\sqrt{2}$ as expected, but the polariton line width Γ is reduced although the number of spins coupled to the cavity mode is increased

This narrowing of the polaritonic peaks Γ is at first a counter intuitive phenomena and surprising since for a single spin or an ensemble of homogeneously broadened spins the system decay rate in the strong coupling regime ($\Omega \gg \kappa$ and γ) is given by $\frac{\gamma+\kappa}{2}$ and does not scale with the number of spins in the cavity. Naively one would expect that this holds true also if an inhomogeneous broadened spin ensembles is coupled to the cavity and that the total decoherence rate should be determined by $\frac{\gamma_{\text{inh}}+\kappa}{2}$. This statement is obviously wrong and to understand the origin of this effect an intuitive explanation is given below. A Tavis-Cummings Hamiltonian as shown and discussed in Chap. 3 is diagonalized and the inhomogeneously broadened spin ensemble is modeled by the spectral coupling density profile as $g_\mu = \Omega\sqrt{\rho(\omega_\mu)/\sum_l \rho(\omega_l)}$. For the sake of simplicity a Gaussian line shape $\rho(\omega)$ of width of γ_{inh} is used and the collective coupling strength Ω is varied. Nevertheless this argument holds to be true also for a convolution of a Lorentzian and Gaussian line shape.

In Fig. 6.2 the eigenvalue spectrum is shown for $\Omega = 0$ and $\Omega = \gamma_{\text{inh}}/10$. The spin ensemble with central frequency ω_s is tuned across the cavity resonance ω_c. In the calculated eigenenergy spectrum the color gradient indicates if an corresponding eigenstate is cavity or spin like. In the case of $\Omega = 0$ a straight *magenta* line corresponding to the uncoupled cavity with eigenfrequency ω_c crosses a bath of *cyan* uncoupled spin states. If the collective coupling strength is now increased to $\Omega = \gamma_{inh}/10$ we observe a distortion in the eigenenergy spectra for the region where the spin ensemble and cavity $\omega_s = \omega_c$ become resonant.

This distortion is a hall mark of the bad cavity regime which is that the broad spin ensemble is weakly coupled to the narrow cavity mode. In the insets of Fig. 6.2 the

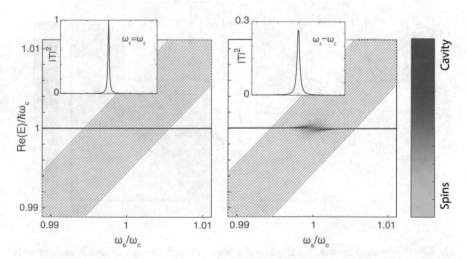

Fig. 6.2 Eigenenergy spectra of a single mode cavity coupled to an ensemble of inhomogeneously broadened γ_{ingh} spins. If an eigenstate is cavity or spin like is indicated by the *color* gradient. (*left*) For $\Omega = 0$ the cavity and spin ensemble remain uncoupled. *Right* If the coupling strength is slightly increased $\Omega = \gamma_{inh}/10$ the system is in the weak coupling regime and a distortion of the eigenenergy is visible. In the (*insets*) the corresponding spectral cavity transmission for $\omega_c = \omega_s$ is plotted

transmission $|T|^2$ spectrum on resonance ($\omega_s = \omega_c$) for the corresponding collective coupling strengths ($\Omega = 0$ and $\Omega = \gamma_{inh}/10$) with dissipation rates $\kappa = \gamma_{inh}/20$, $\gamma = \gamma_{inh}/6$ and probe frequency ω_p is shown. The bare uncoupled cavity ($\Omega = 0$) shows unitary transmission and as the collective coupling strength $\Omega = \gamma_{inh}/10$ is increased the transmission amplitude drops and simultaneously the cavity linewidth is broadened. Such a behavior is expected since the system is in the weak coupling regime the spin ensemble opens up a new dissipation channel for the cavity mode. This result can be understood due to the Purcell effect [15] in which the probability of spontaneous emission of excited spins in the ensemble is increased by the interaction with the electromagnetic field of the cavity. This also has the effect of an enhanced absorption and damping of a cavity mode and explains the broadened cavity line width.

As the collective coupling strength Ω is increased the coupled system enters the strong coupling regime. To show this behavior a Tavis-Cummings Hamiltonian is diagonalized for the following coupling parameters $\Omega = \gamma_{inh}/4$ and $\Omega = \gamma_{inh}$, which are sufficiently strong enough to bring the system into the strong coupling regime. The result is shown in Fig. 6.3 in which two prominent polariton modes are visible in a bath of subradiant spin states. In the *(insets)* of Fig. 6.3 the corresponding cavity transmission for $\omega_c = \omega_s$ and $\kappa = \gamma_{inh}/20$ and $\gamma = \gamma_{inh}/6$ is shown. Since $\Omega \gg \kappa, \gamma$ the system hybridizes and is in the strong coupling regime for both coupling strengths, but in the transmission spectrum the line narrowing of the polariton modes is clearly visible which we identify as the "cavity protection" effect.

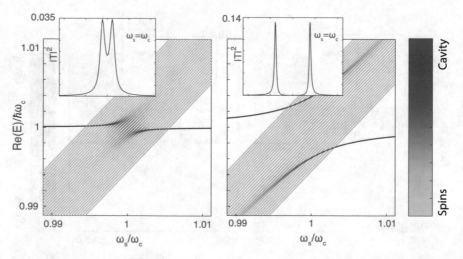

Fig. 6.3 Eigenenergy spectra of a single mode cavity strongly coupled to an inhomogeneously broadened ensemble of spins. Whether an eigenstate is cavity or spin like is indicated by the *color* gradient. *Left* A collective coupling strength $\Omega = \gamma_{inh}/4$ is sufficient to bring the system into the strong coupling regime ($\Omega \gg \kappa = \gamma_{inh}/20$, $\gamma = \gamma_{inh}/6$). *Right* If the collective coupling strength is increased further to $\Omega = \gamma_{inh}/4$ the line narrowing for the polariton modes is clearly visible in the *insets*

In the strong coupling regime the cavity mode couples to a superradiant spin wave and hybridizes to two polariton modes $|\pm\rangle$, which are split by the Rabi frequency Ω_R. The polariton modes and the remaining bath of subradiant states are not entirely decoupled in the presence of inhomogeneous broadening. This spectral overlap of the polariton modes and bath of subradiant states depends on the Rabi splitting as the spectral density of subradiant states follows the spectral spin distribution $\rho(\omega)$. As the energetic gap $\hbar\Omega_R$ is increased an energetic decoupling of the superradiant spin state from the bath of subradiant spin states occurs. This is the basic principle which suppresses damping of the polariton modes induced by spin dephasing. In the limit of large collective coupling strengths the total decoherence rate Γ will then follow the natural limit given by $\Gamma \approx \frac{\kappa+\gamma}{2}$ as will be shown in the latter. This energetic decoupling is only possible in the presence of inhomogeneous line broadening where the wings of the spectral spin distribution $\rho(\omega)$ fall off faster than $1/\omega^2$, a circumstance only valid for non Lorentzian line shapes of $\rho(\omega)$.

6.3 Experimental Verification

In order to demonstrate this behavior experimentally the decay rates Γ of the coupled cavity spin system have to be measured depending on the collective coupling strength Ω. This is possible by applying a microwave pulse sufficiently long to drive the system into a stationary state and the decaying intensity transmitted through the cavity is monitored after the pulse is switched off. In the left panel of Fig. 6.4 the bare cavity response is shown when the spin ensemble is largely detuned from the cavity, which correspond to the case when $\Omega = 0$ omitting the dispersive cavity shift. In the case of having the spin ensemble and cavity in resonance and applying a resonant drive pulse ($\omega_s = \omega_c = \omega_p$) Rabi oscillations are visible after the drive is switched off, as shown in the right panel of Fig. 6.4. The observed decay rates of $\kappa/2\pi = 0.44 \pm 0.1$ MHz (HWHM) and $\Gamma/2\pi = 3 \pm 0.1$ MHz (FWHM) for the bare cavity and hybridized

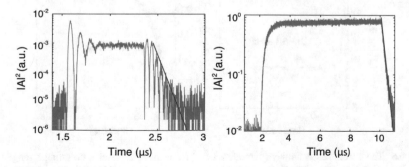

Fig. 6.4 *Left* Resonant cavity transmission ($\omega_p = \omega_c \neq \omega_s$) when the spin ensemble is largely detuned. *Right* Transmission signal and Rabi oscillations when the spin ensemble is shifted into resonance with the cavity ($\omega_p = \omega_c = \omega_s$)

system, respectively, are estimated by applying an exponential fit to the decaying transmitted intensity $|A|^2$.

The collectively enhanced coupling strength Ω has to be scanned in order to determine the scaling behavior of the total decoherence rate Γ. This scaling, i.e. the "cavity protection" effect, can be tested by gradually bleaching or thermalizing the spin ensemble which reduces the effective spin polarization. This corresponds to a reduced number of spins N coupling to the cavity mode, while the shape $\rho(\omega)$ and width γ_{inh} of the inhomogeneous spin distribution remain constant. Such a spin saturation can be achieved by repeatedly probing the cavity with intensities and long microwave pulses, corresponding to $\sim 2.5 \times 10^{-5}$ microwave photons per spin on average in the cavity. This measurement power is already strong enough to slowly excite a non-negligible number of NV spins and leads to a reduced number of ground-state spins coupled to the cavity mode. This process happens on a very slow time scale given by the Purcell shortened spin lifetime. Therefore the system response and transmission can be easily monitored during this procedure from which the decay rate Γ is estimated. This is possible since saturated spins will relax into their ground state on a very long timescale $\propto T_1$ compared to the system decay rate $1/\Gamma$. In Fig. 6.5 such a measurement protocol is implemented and the cavity response for four different coupling strengths Ω is shown.

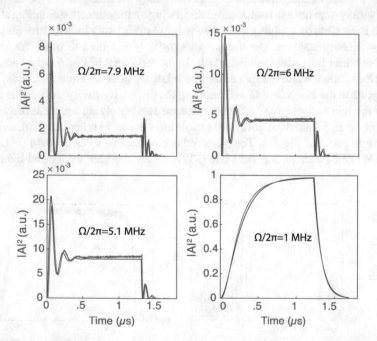

Fig. 6.5 The ensembles is slowly thermalized and reduced the collective coupling strengths Ω. The transmitted intensity through the cavity $|A(t)|^2$ is indicates that by the changed transmission level and decay rate. The experimental data (*blue curve*) is modeled by a full numerical model (*red curve*) from which the collective coupling strength Ω is reffed from (see [9])

The decay rate Γ is then measured for a wide range of coupling strengths until the spin ensemble is entirely bleached. Measuring the decay rate and the coupling strength $\Omega \approx \Omega_R/2$ is somewhat straight forward as long as Rabi oscillations are directly visible, but as soon as the system enters the weak coupling regime it is difficult to directly measure these quantities. In order to precisely estimate all crucial system parameters the dynamical response is modeled by setting up a Tavis-Cummings Hamiltonian as is discussed in Sect. 5.1. In the linear regime a set of first order differential equations can be derived for the expectation values of the cavity operator $\langle a \rangle$ and spin operators $\langle \sigma_j \rangle$, with κ and γ for the cavity and spin dissipation rates, respectively. As is shown in Sect. 5.2.1 and in great detail in Reference [9]. To accurately describe the system dynamics the spectral spin distribution [7, 8] $\rho(\omega) = \sum_j g_j^2 \delta(\omega - \omega_j)/\Omega^2$ has to be included with great care. As shown in [9, 14] the dynamics of the coupled system then can be modeled with great accuracy by setting up a Volterra integral equation $A(t) = \int_0^t d\tau \int d\omega \, \mathcal{K}(\rho(\omega); t - \tau) A(\tau) + \mathcal{F}(t)$, for the cavity amplitude $A(t) = \langle a(t) \rangle$. This includes a memory kernel $\mathcal{K}(t - \tau)$ responsible for the non-Markovian feedback of the NV ensemble on the cavity and the function $\mathcal{F}(t)$ which describes the contribution from an external drive and the initial spin excitation. The numerically computed cavity amplitude $|A(t)|^2$ is compared directly to its experimental counterpart, the intensity transmitted through the cavity, from which the collective coupling strength Ω in the experimental is referred.

The decay rate $\Gamma(\Omega)$ of the coupled system is measured and varies by almost one order of magnitude in a strongly non-monotonic fashion. In the weak-coupling regime the decay rate Γ increases with growing coupling strength Ω due to the Purcell-effect [15] as the cavity mode increasingly couples to the spin ensemble. Entering the strong-coupling regime, this trend reverses and Γ decreases with growing Ω, as shown in Fig. 6.6 where both decay rates $\Gamma(\Omega)$ derived experimentally and numerically are depicted. While the maximally reachable value of Ω in the experiment already leads to a considerable reduction of Γ by 50% below its maximum, a further reduction

Fig. 6.6 Observed decay rates $\Gamma(\Omega)$ by the experiment *black squares* and full numerical simulation *red squares*. Analytical decay rates for homogeneous Γ_{hom} (*green*) and inhomogeneous Γ_{inh} (*magenta*) broadened spin ensembles are plotted as well for parameters $\kappa/2\pi = 0.44$ MHz (FWHM), $\gamma \to 0$ and $\Omega/2\pi = 8.6$ MHz

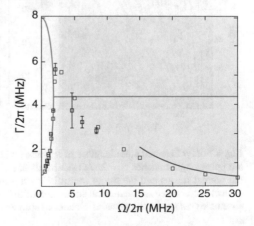

of the decay rate with increasing coupling strength by one order of magnitude is possible. The natural limit for the system total decay rate is then given again by $\Gamma \approx \frac{\kappa + \gamma}{2}$ as discussed earlier. The observed total decay rates are compared to the analytically expression of the system decay rate of a homogeneous Lorentzian spin distribution coupled to the single mode cavity

$$\Gamma_{\text{hom}} = \frac{\kappa + \gamma \pm \text{Re}(\sqrt{(\gamma - \kappa)^2 - 4\Omega^2})}{2} \tag{6.1}$$

which was discussed in Sect. 3.1.3. A similar analytical expression for the total decay rate can be found in the limit of large coupling strength Ω by performing a Laplace transform of the Volterra equation which reads [7, 16]

$$\Gamma_{\text{inh}} = \frac{\kappa + \gamma + 2\pi\Omega^2 \rho(\omega_s \pm \Omega)}{2}. \tag{6.2}$$

This shows that only for rather large collective coupling strengths Ω the system decay rate will be determined by the natural limit of $\Gamma = \frac{\kappa + \gamma}{2}$.

The cavity protection effect is also beneficial for the realization of coherent control schemes. To demonstrate this experimentally a modulated driving tone is injected into the cavity which has frequency components at $\omega_p \pm \Omega_R/2$. This probe tones are consequently in resonance with the polariton modes, i.e. both normal modes. This is achieved by using a rectangular pulse train in which every pulse has a length given by the Rabi frequency and each consecutive pulse has a π phase shift, as is shown in Fig. 6.7. Since the cavity transmission is maximal for such a modulated drive signal large intra cavity fields strength are created which allows strong driving of the spin

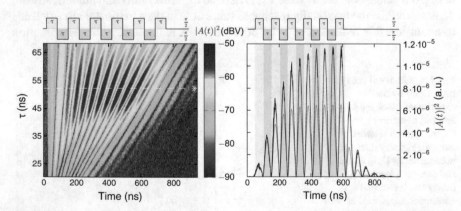

Fig. 6.7 *Left* Scanned pulse length τ of π phase switched consecutive rectangular pulses. *Right* If the pulse length matches $\tau = 2\pi/\Omega$ sidebands at $\omega_p \pm \Omega$ are created which are in resoance with the normal modes and the cavity transmission is substantially increased. The experimental data (*black*) is compared to a numerical simulation of the system for an inhomogeneous broadened spin ensemble (*red*) and homogeneous broadened spin ensemble (*green*)

ensemble. Such a measurement protocol is implemented and shown in Fig. 6.7 by probing the cavity with a driving field with a carrier frequency $\omega_p = \omega_c = \omega_s$ and a periodical modulation with tunable period τ. The giant oscillations in the cavity transmission corresponding to a coherent stimulus of the coupled system and Rabi oscillation are observed after the drive is switched off. A maximum oscillation amplitude occurs exactly at the point where the modulation period τ coincides with the inverse of the Rabi frequency $2\pi/\Omega_R$. The stationary state of the oscillating amplitude exceeds the steady state amplitude by two orders of magnitude compared to driving the system with a rectangular pulse with $\omega_c = \omega_s = \omega_p$. Although the net power applied at the cavity input is exactly the same in both cases. For comparison with the experimental data, numerical results [9, 14] for an inhomogeneously broadened spin ensemble with a q-Gaussian as well as for a homogeneous broadened Lorentzian spin-density are shown in Fig. 6.7. This clearly shows the substantially lower excitation amplitudes for a homogeneously broadened spin ensemble, and indicates potential of the "cavity protection effect" for the implementation of coherent control schemes.

6.4 Conclusion

The cavity protection effect is an intrinsic mechanism which reduces the total decoherence rate of the system. The suppression of spin dephasing is therefore possible by strongly coupling the ensemble to the cavity mode. Although it is rather simple mechanism and follows directly from a very intuitive picture it has been predicted [6] and experimentally demonstrated [14] only very recently. This is due to the fact that in standard cavity QED experiments ensembles of atoms in optical cavities are normally less or almost not prone to inhomogeneous line broadening [17]. However, as solid-state cavity QED systems become of general interest for future quantum technologies this effect can become very important for the improvement of the coherence times of quantum memories. This can be seen as a first proof of principle and how potential disadvantages can be overcome naturally by hybridization. This also means many more beautiful and complex collective spin phenomena can be observed in future solid-state cavity QED experiments although inhomogeneous spin broadening is present.

References

1. Y. Kubo, F.R. Ong, P. Bertet, D. Vion, V. Jacques, D. Zheng, A. Dréau, J.-F. Roch, A. Auffeves, F. Jelezko, J. Wrachtrup, M.F. Barthe, P. Bergonzo, D. Esteve, Strong coupling of a spin ensemble to a superconducting resonator. Phys. Rev. Lett. **105**(14), 140502 (September 2010)
2. D.I. Schuster, A.P. Sears, E. Ginossar, L. DiCarlo, L. Frunzio, J.J.L. Morton, H. Wu, G.A.D. Briggs, B.B. Buckley, D.D. Awschalom, R.J. Schoelkopf, High-cooperativity coupling of electron-spin ensembles to superconducting cavities. Phys. Rev. Lett. **105**(14), 140501 (September 2010)

3. R. Amsüss, Ch. Koller, T. Nöbauer, S. Putz, S. Rotter, K. Sandner, S. Schneider, M. Schramböck, G. Steinhauser, H. Ritsch, J. Schmiedmayer, J. Majer, Cavity QED with magnetically coupled collective spin states. Phys. Rev. Lett. **107**(6), 060502 (August 2011)

4. S. Probst, H. Rotzinger, S. Wünsch, P. Jung, M. Jerger, M. Siegel, A.V. Ustinov, P.A. Bushev, Anisotropic rare-earth spin ensemble strongly coupled to a superconducting resonator. Phys. Rev. Lett. **110**(15), 157001 (April 2013)

5. C.W. Zollitsch, K. Mueller, D.P. Franke, S.T.B. Goennenwein, M.S. Brandt, R. Gross, H. Huebl, High cooperativity coupling between a phosphorus donor spin ensemble and a superconducting microwave resonator. Appl. Phys. Lett. **107**(14), 142105 (October 2015). ISSN 0003-6951, 1077-3118

6. Z. Kurucz, J.H. Wesenberg, K. Mølmer, Spectroscopic properties of inhomogeneously broadened spin ensembles in a cavity. Phys. Rev. A **83**(5), 053852 (May 2011)

7. I. Diniz, S. Portolan, R. Ferreira, J.M. Gérard, P. Bertet, A. Auffèves, Strongly coupling a cavity to inhomogeneous ensembles of emitters: Potential for long-lived solid-state quantum memories. Phys. Rev. A **84**(6), 063810 (December 2011)

8. K. Sandner, H. Ritsch, R. Amsüss, Ch. Koller, T. Nöbauer, S. Putz, J. Schmiedmayer, J. Majer, Strong magnetic coupling of an inhomogeneous nitrogen-vacancy ensemble to a cavity. Phys. Rev. A **85**(5), 053806 (May 2012)

9. O. Dmitry, Krimer, Stefan Putz, Johannes Majer, Stefan Rotter, Non-Markovian dynamics of a single-mode cavity strongly coupled to an inhomogeneously broadened spin ensemble. Phys. Rev. A **90**(4), 043852 (October 2014)

10. Xiaobo Zhu, Shiro Saito, Alexander Kemp, Kosuke Kakuyanagi, Shin-ichi Karimoto, Hayato Nakano, William J. Munro, Yasuhiro Tokura, Mark Everitt, Kae S. Nemoto, Makoto Kasu, Norikazu Mizuochi, Kouichi Semba, Coherent coupling of a superconducting flux qubit to an electron spin ensemble in diamond. Nature **478**, 221–224 (September 2011)

11. Y. Kubo, I. Diniz, A. Dewes, V. Jacques, A. Dréau, J.-F. Roch, A. Auffeves, D. Vion, D. Esteve, P. Bertet, Storage and retrieval of a microwave field in a spin ensemble. Phys. Rev. A **85**(1), 012333 (January 2012)

12. C. Grèzes, B. Julsgaard, Y. Kubo, M. Stern, T. Umeda, J. Isoya, H. Sumiya, H. Abe, S. Onoda, T. Ohshima, V. Jacques, J. Esteve, D. Vion, D. Esteve, K. Mølmer, B.P. Multi-mode storage and retrieval of microwave fields in a spin ensemble, (2014) arXiv:1401.7939

13. Wu Hua, Richard E. George, Janus H. Wesenberg, Klaus Mølmer, David I. Schuster, Robert J. Schoelkopf, Kohei M. Itoh, Arzhang Ardavan, John J. L. Morton, G. Andrew, D. Briggs, Storage of multiple coherent microwave excitations in an electron spin ensemble. Phys. Rev, Lett. **105**(14), 140503 (September 2010)

14. S. Putz, D.O. Krimer, R. Amsüss, A. Valookaran, T. Nöbauer, J. Schmiedmayer, S. Rotter, J. Majer, Protecting a spin ensemble against decoherence in the strong-coupling regime of cavity QED. Nat. Phys. **10**(10):720–724, (October 2014). ISSN 1745-2473

15. E.M. Purcell, Proceedings of the American Physical Society. Phys. Rev. **69**(11–12), 681 (June 1946)

16. D.O. Krimer, B. Hartl, S. Rotter, Hybrid quantum systems with collectively coupled spin states: suppression of decoherence through spectral hole burning. Phys. Rev. Lett. **115**(3), 033601 (July 2015)

17. G. Rempe, R.J. Thompson, R.J. Brecha, W.D. Lee, H.J. Kimble, Optical bistability and photon statistics in cavity quantum electrodynamics. Phys. Rev. Lett. **67**(13), 1727–1730 (September 1991)

Chapter 7
Engineering of Long-Lived Collective DarkStates—Spectral Hole Burning

This chapter is based on the following research paper:

- S. Putz, A. Angerer, D. O. Krimer, R. Glattauer, W. J. Munro, S. Rotter, H.-J. Schmiedmayer, J. Majer, **"Spectral hole burning and its application in microwave photonics"** Nature Photonics, **11** (2017) (1), 36–39

and uses parts of it.

7.1 Introduction

In the previous chapter is was shown how the total decoherence Γ in a hybrid solid-state cavity QED system scales. The spin dephasing introduced by an inhomogeneous broadened spin ensemble can be suppressed by energetically decoupling the polariton modes $|\pm\rangle$ from a bath of subradiant states. Nevertheless, only for rather large collectively enhanced coupling strengths Ω the natural limit of $\Gamma = \frac{\kappa + \gamma}{2}$ is reachable. In this regime an excitation in $|\pm\rangle$ is shared equally between the cavity mode and a superradiant spin state. Hence the total decoherence is given by the mean of the cavity decay rate κ and single spin dissipation rate γ of the system. Although the spin dephasing can be suppressed by the "cavity protection effect", the cavity interface still induces considerable dissipation which limits the total system coherence time. In this chapter I will discuss how it is possible to engineer long-lived polariton modes which live mostly in the spins and only very little in the cavity mode by spectral hole burning of the spin ensemble. This technique drastically reduces the system coherence time (since $\kappa \gg \gamma$) and is able to suppress spin dephasing and dissipation from a cavity interface as theoretical predictions suggest [1, 2]. Here we identify this states as long-lived collective dark states similar to what is shown in [3, 4]. Moreover the observed coherence times truly live up to the promise of hybrid devices: that the combined device performs better than its individual subcomponents. To my best knowledge this is the first demonstration of the full potential off hybridization in an all solid-state device.

© Springer International Publishing AG 2017
S. Putz, *Circuit Cavity QED with Macroscopic Solid-State Spin Ensembles*,
Springer Theses, DOI 10.1007/978-3-319-66447-7_7

7.1.1   The Principle of Spectral Hole Burning

Spectral hole burning is a technique widely used in solid state spectroscopy and has found many application in past and present experiments [5–10]. In a simple hole burning spectroscopy measurement the absorption spectra of an ensemble of broadened spins or atoms with a linewidth γ_{inh} is monitored. With a sharp high intensity pulse spins in a narrow frequency window are then excited and saturated, leaving a small hole in the absorption spectra, i.e. spectral hole burning. This hole represents a spin packet within the ensemble which can be understood as the smallest observable homogeneously broadened spin linewidth γ as is schematically depicted in Fig. 7.1.

Spectral hole burning further allows the selection of spins in a well defined frequency window by pumping them into an eventually existing long-lived auxiliary state. This method is used to implement atomic frequency combs and has proven very successful in optical quantum memory protocols [11–13] or for testing macroscopic entanglements [14]. A similar approach can be made also if a third level is not available to shelf away distinct spin components. Spectral bleaching by saturating spins in a narrow frequency window brings some two-level systems into a mixture of their ground and excited state and as a result their effective dipole moment averages to zero as depicted in Fig. 7.1. In a cavity-spin system this modifies the spectral coupling density profile $g_{\mu} = \Omega \sqrt{\rho(\omega_{\mu}) / \sum_l \rho(\omega_l)}$ of the ensemble and one can arbitrarily shape the spectral spin distribution coupled to the cavity mode. If spins are excited and in a 50:50 mixture they can no longer couple to the cavity mode, but eventually decay towards their ground state by emitting a photon into the cavity. The time scale on which this progress occurs is given by the single spin cavity interaction strength. This rate will be generally higher than the inverse spin life time $1/T_1$ due to the Purcell effect, but due to the weak cavity spin interaction strengths this time scale can still be sufficiently long to coherently manipulate the system. Therefore spectral ensemble shaping enables us to engineer long-lived collective dark states which can be followed by an arbitrary protocol as discussed in the latter.

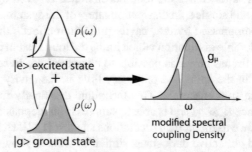

Fig. 7.1 The principle of spectral hole burning: spectral spin components are saturated and bleached. One can arbitrarily modify the spectral coupling density profile when spins in a narrow frequency window are brought into a mixture of their ground and excited state

7.1.2 Collective Spin Dark States

As shown in Chap. 3 dark states are a natural result of coupling an ensemble of two-level systems to the cavity mode. In the absence of inhomogeneous spectral broadening this subradiant states are entirely decoupled from the cavity mode and therefore remain entirely dark [15]. In large ensembles of spins it is common to use to the Dicke model [16] and to refer to super- and subradiant spin states rather than bright and dark states, respectively. In the presence of inhomogeneous spectral spin broadening subradiant states become non degenerate, and are not entirely decoupled from the polariton modes as was shown in Chap. 6. This bath of subradiant states is in general a source of decoherence and accelerates the evolution of a spin superradiant state into a subradiant and almost uncoupled spin state. However, this bath of subradiant states can be modified by spectral hole burning, which gives rise to very unique subradiant spin states to which we refer as collective dark states in the later. The spectral hole burning bleaches spectral parts of the spin ensemble and modifies the coupling distribution g_μ. The coupling of the remaining parts of the spin ensemble gives rise to collective dark states which are isolated from the bath of subradiant states and the cavity mode. Therefore a substantial suppression of the decoherence rate is possible beyond the natural limit attainable by the cavity protection effect.

For an intuitive understanding of this effect we discretize and solve the eigenvalue problem of the single mode cavity coupled to an inhomogeneously broadened spin ensemble as is shown in Sect. 3.3. Two optimal positions for burning narrow spectral holes in the spectral spin distribution are found. If a spectral hole of width Δ is burned at the center of the spin ensemble ω_s or at the position of the polariton modes $\omega_s \pm \Omega_R/2$ an anti-symmetric superposition of subradiant states, namely a dark state, isolated from the bath of subradiant states emerges. In a simplified picture these states can be written as spins blue and red detuned with respect to the cavity mode and spectral hole. For such an anti-symmetric spin state $|A\rangle \approx (\downarrow_B\uparrow_R\rangle - \uparrow_B\downarrow_R\rangle)/\sqrt{2}$ the coupling to the cavity mode

$$D\rangle \approx \frac{1}{\sqrt{\Delta^2 + 2g_\mu^2}} \left(g_\mu \sqrt{2} A\rangle 0\rangle_c + \Delta \downarrow_B\downarrow_R\rangle 1\rangle_c \right),$$

is directly proportional to the width of the spectral hole Δ with an effective cavity spin coupling strength g_μ. Such an engineered dark state $|D\rangle$ results in a peak in the transmission signal and can be substantially narrower than cavity linewidth κ. The linewidth $\Gamma_D \geq \gamma$ is directly related to the width of the spectral hole , therefore the cavity and spin components of the dark state $|D\rangle$ can be controlled directly by Δ. Such states can be produced only in the center of the spectral spin distribution and in the proximity of the polariton modes. Due to the large collective coupling strength Ω spectral holes at different spectral positions would create dark states that hybridize and are not located in the center of the created spectral holes (Fig. 7.2).

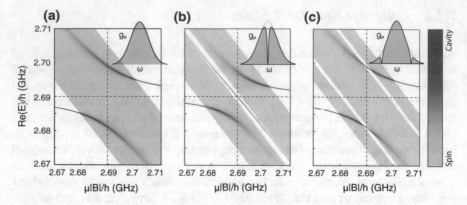

Fig. 7.2 Eigenenergy spectrum for three different spectral coupling density profiles shown in **a**, **b** and **c**. In the case of the natural broadened spectral spin distribution in **a** two clear polariton modes (*magenta*) are visible in a bath of sub-radiant states (*cyan*). In **b** and **c** the optimal positions for spectral holes are shown where an dark state isolated from the bath of subradiant states emerges

7.2 Experimental Implementation of Spectral Hole Burning

7.2.1 Spectral Hole Burning

Spectral hole burning is implemented by a coherent control scheme as discussed in Chap. 6. The cavity probe tone with carrier frequency $\omega_p = \omega_c = \omega_s$ is modulated by a sinusoidal signal with frequency $\Omega_R/2$ $(\sin(\Omega_R t/2)e^{-i\omega_p t})$ and a Gaussian envelope of width $\Delta/2\pi = 470$ kHz. This pulse produces two sidebands at $\omega_p \pm \Omega_R/2$ with a width Δ, as shown in Fig. 7.3. The drive is in resonance with both polariton

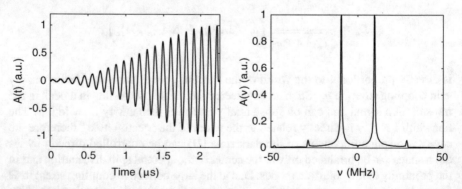

Fig. 7.3 A down converted hole burning pulse sequence use for spectral hole burning. A carrier signal with frequency $\omega_p = \omega_s = \omega_c$ is sine modulated with a frequency $\Omega_R/2$ $(\sin(\Omega_R t/2)e^{-i\omega_p t})$ and a Gaussian envelope. The signal is Fourier transformed and has two side bands at $\omega_p \pm \Omega_R/2$ with a band with of 470 kHz

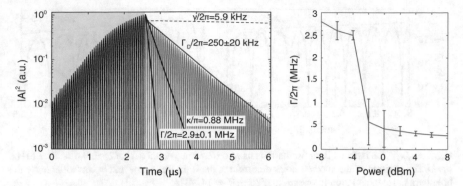

Fig. 7.4 A hole burning pulse ($\Omega_R/2$ ($\sin(\Omega_R t/2)e^{-i\omega_p t}$) (*gray area*) bleaches spins at distinct frequencies creating two spectral holes with a bandwidth of $\Delta/2\pi = 470$ kHz (FWHM) at $\omega_c/2\pi \pm$ 9.6 MHz. After the drive power reaches a threshold (*right*) the decay of Rabi oscillations is slowed down with a minimal rate of $\Gamma_h/2\pi = 250 \pm 20$ kHz (*red*), we also plot the characteristic decay envelopes (*black*) for $\kappa/2\pi = 0.44$ MHz, $\Gamma/2\pi = 2.9$ MHz and $\gamma/2\pi = 5.9$ kHz. The signal decays a factor of 3.5 slower than the pure cavity dissipation rate κ

modes and therefore create large intracavity field amplitudes. Power values of up to 20 milliwatt, corresponding to a steady state of $\approx 10^4$ photons per spin in the cavity, are applied and are strong enough to selectively bleach spectral spin components at frequencies $\omega_s \pm \Omega_R/2$. Saturated spins decay slowly towards their ground state on a time scale of $\gtrsim 10$ ms due to their Purcell shortened spin life time [17] of $T_1 = 45$ s.

The hole burning scheme is optimized by scanning the pulse intensity as is shown in Fig. 7.4. When a power threshold, strong enough to bleach sufficient spectral spin components at distinct frequencies, is reached, isolated long-lived dark states emerge indicated by the slow decay of the hole burning pulse. After the hole burning pulse is switched off, coherent Rabi oscillations occur and the transmitted intensity $|A(t)|^2$ through the cavity decays exceptionally slower than for drive powers below the threshold. The Rabi frequency can be controlled by altering the position of the spectral holes simply by choosing a different modulation frequency of the hole burning pulse, as shown in Fig. 7.5. In the presented experiments the best achievable decay rate is $\Gamma_D/2\pi = 250 \pm 10$ kHz for spectral holes at $\omega_p/2\pi \pm 9.6$ MHz which is one order of magnitude smaller than the decay rate given by the width of the polariton peaks $\Gamma/2\pi = 2.9$ MHz. Most importantly this decay rates is a factor of 3.5 smaller than the bare cavity decay rate $\kappa/2\pi = 0.44$ MHz (HWHM).

7.2.2 Dark State Spectroscopy

The created long-lived dark states can be observed directly in the cavity transmission spectroscopic measurements after the hole burning. The steady state intensity transmitted through the cavity $|A(\omega_p)|^2$ is measured by scanning the probe frequency

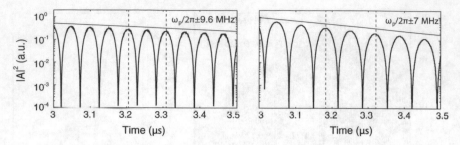

Fig. 7.5 *Left* The induced Rabi oscillations have a frequency equal to $\omega_p/2\pi \pm 9.6 = 19.2$ MHz. *Right* The position of the spectral holes determines the Rabi frequency and is controllable by the hole burn pulse modulation frequency $\omega_p/2\pi \pm 7 = 14$ MHz

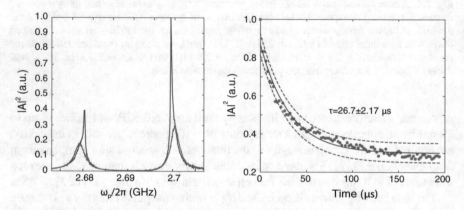

Fig. 7.6 Dark state spectroscopy measurement before (*blue*) and after (*red*) a hole burning pulse has been applied. The spectral holes and dark states $\Gamma_D/2\pi = 440$ kHz (FWHM) were created directly on top of the polariton peaks $\Gamma/2\pi = 2.9$ MHz (FWHM). The life time of the spectral holes $\tau = 26.7 \pm 2.17$ μs is measured by tracing the decaying amplitude of one of the created dark states

ω_p before and ≥ 5 μs after spectral holes were burnt at positions equal to $\pm\Omega_R/2$. The system is probed with low intensities, $<10^{-5}$ photons per spin in the cavity, and two narrow peaks emerge directly on top of the polaritonic peaks as shown in Fig. 7.6. The created holes and narrow peaks in the transmission signal decay with a characteristic time constant $\tau = 26.7 \pm 2.17$ μs due to spin diffusion which limits the spectral hole lifetime in this experiment. However, this time constant τ is a factor of four longer compared to the best achievable spin echo times presented in Chap. 9, $T_2 = 4.8 \pm 1.6$ μs and $T_{1\rho} = 6.4 \pm 0.59$ μs measured by CPMG and stimulated echos, respectively. This also means that the spin dissipation rate $\gamma/2\pi \approx 1/2\pi\tau = 5.9$ kHz is dominated by spin diffusion in our experiment since $T_2 \approx T_{1\rho}$. Although limited by the same process the spectral hole life time is more than a factor or four longer which can be explained by the miss alignment of the external d.c. magnetic field [18] with respect to the NV axis and a bath of excess electron and nuclear spins in the host material.

7.2.3 Dark State Dynamics

The dynamical response of the system is probed in the linear regime in order to demonstrate that these engineered long-lived dark states can be used for the coherent exchange of a low intensity excitation. Therefore the system is probed before and after a hole burning sequence with low probe intensities, $<10^{-5}$ photons per spin in the cavity. A short pulse is applied with a carrier frequency $\omega_p = \omega_c = \omega_s$ and further is sinusoidally modulated with frequency $\Omega_R/2$. At times before the hole burning sequence the unchanged system decay rate $\Gamma/2\pi = 2.9 \pm 0.1$ MHz is observed in the decaying Rabi oscillations, as shown in Fig. 7.7. More than 5 μs after the hole burning sequence and the intra cavity field has decayed the system is probed again by the same pulse sequence. As a first sign of the improved coherence time the driven Rabi oscillations need substantially longer to set into a stationary state. After the drive signal is switched off, two rates are distinguishable in the decaying Rabi oscillations as is shown in Fig. 7.7. At first a reduced decay rate of the polariton modes $\Gamma'/2\pi = 1.1 \pm 0.1$ MHz, corresponding to a reduced width of the polaritonic peaks is observed. The polariton modes have a reduced energetic overlap with the bath of subradiant states due to the created spectral holes, which

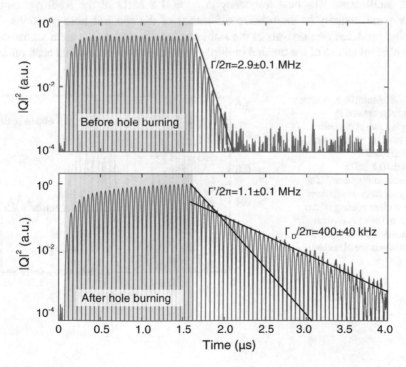

Fig. 7.7 Dynamical response before and after spectral holes were created at $\approx \omega_s \pm \Omega_R/2$. The system is probed with low intensities, $<10^{-5}$ photons per spin, and a $\Omega_R/2$ sinusoidally modulated pulse (*gray area*) with carrier frequency $\omega_p = \omega_c = \omega_s$

explains this first slowed down decay rate. A second drastically slower decay rate of $\Gamma_D/2\pi = 0.4 \pm 0.04$ MHz is observed and attributed to the created long-lived dark states. The fact that two clear distinguishable decay rates are observed is a signature of the dark states decoupled from the polariton modes. Since two dark states are created, symmetrically with respect to the bare cavity frequency, they induce long-lived coherent Rabi oscillations. During which energy is exchanged between the cavity and spin ensemble. As a matter of fact the created dark states include different spin components at frequency of $\omega_s \pm \Omega_R/2$. Hence, during the coherent Rabi oscillations energy is also exchanged between the spin components of both dark states. This means a 4π rotation or two full Rabi periods are needed to transfer energy from the spin components of one dark state into the cavity and back.

A way to point to a future application of this technique is the creation of multiple spectral holes in the proximity of the polariton modes. This allows to create multiple coherent dark states beating against each other as shown in Fig. 7.8. In the experiment four holes are burned by two consecutive sine modulated pulses with modulation frequencies $\nu_1 = \pm 9$ MHz and $\nu_2 = \pm 10.8$ MHz and a Gaussian envelope of 300 kHz bandwidth. The dynamical response after the hole burning sequence is probed by low intensities and a short sine modulated pulse as shown earlier. After the drive is switched off the driving signal exhibits a clear beating in the decaying Rabi oscillations. The beat frequency $\Delta_{\nu_{21}} = 1.8$ MHz of the Rabi oscillations corresponds exactly to the relative difference of the spectral hole positions. The beating produces two revivals in the Rabi oscillations which is a clear signature of the coherent nature of the created multiple dark states beating against each other.

Fig. 7.8 Multiple dark states have been created by applying two consecutive hole burning pulses with $\nu_1 = \pm 9$ and $\nu_2 = \pm 10.8$ MHz modulation frequency. The decaying Rabi oscillations show a clear beating of the coherent dark states with frequency ≈ 1.8 MHz and two revivals are observed

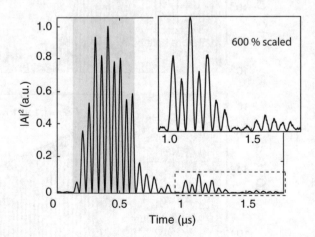

7.3 Conclusion

These results demonstrate that it is certainly possible to engineer long-lived polariton modes. The observed decay rates are substantially improved beyond the limit attainable by the cavity protection effect. The achieved improvement truly lives up to the promise of hybrid systems to perform better than their individual subcomponents. To my best knowledge this is the first proof of this promise and the implementation of spectral hole burning in an all microwave hybrid solid-states device. In the presented experiments the life time of the spectral holes is limited by the single spin-cavity interaction strength. However, for NV spins also a third auxiliary level is available which could also be used to create even longer lived spectral holes. If a spectral hole is created by shelving the spin population into the off resonant third NV level the created spectral hole life time is limited only by the extremely long T_1 time. This would allow an almost persistent suppression of spin dephasing and dissipation introduced by the cavity interface.

A first key step towards an on-chip all solid-state microwave frequency comb with long coherence times is the creation of multiple spectral holes and dark states. This approach may be a key component in future quantum memories and transducer networks. This demonstrates further that it is possible to fully access the "decoherence free" [19] subspace in the experiment and the selectively preparation of protected states [20] could become possible to suppress decoherence even further. Therefore, this spectral hole burning technique opens up the way for long-lived quantum memories, solid-state microwave frequency combs [13], optical-to-microwave quantum transducers [21, 22] and spin squeezed states [23] in circuit cavity QED experiments. One should note that the presented technique is a somewhat different approach compared to cavity line narrowing [24, 25] and the existence of dark states known from experiments employing electrical induced transparency (EIT) [11, 26, 27]. The presented technique could become an additional powerful method in quantum computation devices using atom or spin ensembles and the presented spectral hole burning technique is gives rise to implement many new protocols.

References

1. D.O. Krimer, B. Hartl, S. Rotter, Hybrid quantum systems with collectively coupled spin states: Suppression of decoherence through spectral hole burning. Phys. Rev. Lett. **115**(3), 033601 (July 2015)
2. G. Bensky, D. Petrosyan, J. Majer, J. Schmiedmayer, G. Kurizki, Optimizing inhomogeneous spin ensembles for quantum memory. Phys. Rev. A **86**(1), 012310 (July 2012)
3. X. Zhu, Y. Matsuzaki, R. Amsüss, K. Kakuyanagi, T. Shimo-Oka, N. Mizuochi, K. Nemoto, K. Semba, W.J. Munro, S. Saito, Observation of dark states in a superconductor diamond quantum hybrid system. Nat. Commun. **5** (April 2014)
4. X. Zhang, C.-L. Zou, N. Zhu, F. Marquardt, L. Jiang, H.X. Tang, Magnon dark modes and gradient memory. Nat. Commun. **6**, 8914 (Nov 2015)

5. G. Feher, Electron spin resonance experiments on donors in silicon. I. Electronic structure of donors by the electron nuclear double resonance technique. Phys. Rev. **114**(5), 1219–1244 (June 1959)

6. P.W. Anderson, Absence of diffusion in certain random lattices. Phys. Rev. **109**(5), 1492–1505 (Mar 1958)

7. R.T. Harley, M.J. Henderson, R. M, Macfarlane, Persistent spectral hole burning of colour centres in diamond. J. Phys. C Solid State Phys. **17**(8), L233 (1984). ISSN 0022-3719

8. S. Volker, Hole-burning spectroscopy. Annu. Rev. Phys. Chem. **40**(1), 499–530 (1989)

9. M. Nilsson, L. Rippe, S. Kröll, R. Klieber, D. Suter, Hole-burning techniques for isolation and study of individual hyperfine transitions in inhomogeneously broadened solids demonstrated in pr^{3+}:y_2sio_5. Phys. Rev. B **70**(21), 214116 (Dec 2004)

10. R. Purchase, S.Völker, Spectral hole burning: Examples from photosynthesis. Photosynth. Res. **101**(2–3), 245–266 (Sept 2009). ISSN 0166-8595

11. M. Fleischhauer, M.D. Lukin, Quantum memory for photons: Dark-state polaritons. Phys. Rev. A **65**(2), 022314 (Jan 2002)

12. M. Afzelius, I. Usmani, A. Amari, B. Lauritzen, A. Walther, C. Simon, N. Sangouard, J. Minář, H. de Riedmatten, N. Gisin, S. Kröll, Demonstration of atomic frequency comb memory for light with spin-wave storage. Phys. Rev. Lett. **104**(4), 040503 (Jan 2010)

13. de H. Riedmatten, M. Afzelius, M.U. Staudt, C. Simon, N. Gisin, A solid-state light–matter interface at the single-photon level. Nature **456**(7223), 773–777 (Dec 2008). ISSN 0028-0836

14. I. Usmani, C. Clausen, F. Bussières, N. Sangouard, M. Afzelius, N. Gisin, Heralded quantum entanglement between two crystals. Nat. Photonics **6**(4), 234–237 (April 2012). ISSN 1749-4885

15. S. Filipp, M. Göppl, J.M. Fink, M. Baur, R. Bianchetti, L. Steffen, A. Wallraff, Multimode mediated qubit-qubit coupling and dark-state symmetries in circuit quantum electrodynamics. Phys. Rev. A **83**(6), 063827 (June 2011)

16. R.H. Dicke, Coherence in spontaneous radiation processes. Phys. Rev. **93**(1), 99–110 (Jan 1954)

17. R. Amsüss, Ch. Koller, T. Nöbauer, S. Putz, S. Rotter, K. Sandner, S. Schneider, M. Schramböck, G. Steinhauser, H. Ritsch, J. Schmiedmayer, J. Majer, Cavity QED with magnetically coupled collective spin states. Phys. Rev. Lett. **107**(6), 060502 (Aug 2011)

18. P.L. Stanwix, L.M. Pham, J.R. Maze, D. Le Sage, T.K. Yeung, P. Cappellaro, P.R. Hemmer, A. Yacoby, M.D. Lukin, R.L. Walsworth, Coherence of nitrogen-vacancy electronic spin ensembles in diamond. Phys. Rev. B **82**(20), 201201 (Nov 2010)

19. A. Beige, D. Braun, P.L. Knight, Driving atoms into decoherence-free states. New J. Phys. **2**(1), 22 (Sept 2000). ISSN 1367-2630

20. D. Plankensteiner, L. Ostermann, H. Ritsch, C. Genes, Selective protected state preparation of coupled dissipative quantum emitters. Sci. Rep. **5**, 16231, (Nov 2015). ISSN 2045-2322

21. K. Stannigel, P. Rabl, A.S. Sørensen, P. Zoller, M.D. Lukin, Optomechanical transducers for long-distance quantum communication. Phys. Rev. Lett. **105**(22), 220501 (Nov 2010)

22. S. Blum, C. O'Brien, N. Lauk, P. Bushev, M. Fleischhauer, G. Morigi, Interfacing microwave qubits and optical photons via spin ensembles. Phys. Rev. A **91**(3), 033834 (Mar 2015)

23. D.J. Wineland, J.J. Bollinger, W.M. Itano, D.J. Heinzen, Squeezed atomic states and projection noise in spectroscopy. Phys. Rev. A **50**(1), 67–88 (July 1994)

24. G. Hernandez, J. Zhang, Y. Zhu, Vacuum Rabi splitting and intracavity dark state in a cavity-atom system. Phys. Rev. A **76**(5), 053814 (Nov 2007)

25. M. Sabooni, Q. Li, L. Rippe, R.K. Mohan, S. Kröll, Spectral engineering of slow light, cavityline narrowing, and pulse compression. Phys. Rev. Lett. **111**(18), 183602 (Oct 2013)

26. M. Fleischhauer, A. İmamoğlu, J.P. Marangos, Electromagnetically induced transparency: Optics in coherent media. Rev. Mod. Phys. **77**(2), 633–673 (July 2005)

27. M.D. Lukin, M. Fleischhauer, M.O. Scully, V.L. Velichansky, Intracavity electromagnetically induced transparency. Opt. Lett. **23**(4), 295 (Feb 1998). ISSN 0146-9592, 1539-4794

Chapter 8
Amplitude Bistability with Inhomogeneous Spin Broadening—Driven Tavis-Cummings

This chapter is a first set of experiments in connection with the following research article:

- Andreas Angerer, <u>Stefan Putz</u>, Dmitry O. Krimer, Thomas Astner, Matthias Zens, Ralph Glattauer, Kirill Streltsov, William J Munro, Kae Nemoto, Stefan Rotter, Jörg Schmiedmayer, Johannes Majer, **"Dynamical Exploration of Amplitude Bistability in Engineered Quantum Systems"** arXiv preprint arXiv:1703.04779 (2017)

8.1 Introduction

In the previous chapters it was shown that the strong coupling regime of cavity QED is reachable by collectively enhanced spin cavity interactions, how the total coherence rate scales with the collective interaction and how the coherence times can be improved by spectral hole burning. Nevertheless, the strongly coupled system was studied and modeled mainly in the linear regime, and only few non linear effects were observed revealing the anharmonicity of the coupled spin system. A way to observe the spin two-level nature in the presented experiments is to increase the driving power and to bring the system into a regime where the description by a collection of coupled harmonic oscillators fails. A bistable behavior eventually occurs in the high cooperativity regime and are a direct signature of the spin two-level systems coupled to the cavity mode. This bifurcation gives rise to a phenomenon known as "optical" bistability [1]. There has been put much effort in experiment and theoretical concepts behind amplitude bistability [2–7] which was belived to find applications in optical computing devices [8, 9]. To my best knowledge this is the first implementation and demonstration of amplitude bistability in a solid-state microwave hybrid quantum system. This devices give further the opportunity to observe dynamical bifurcation processes which could not been resolved previously due to short spin or atom life times.

© Springer International Publishing AG 2017
S. Putz, *Circuit Cavity QED with Macroscopic Solid-State Spin Ensembles*,
Springer Theses, DOI 10.1007/978-3-319-66447-7_8

8.2   The Principle of Bistability

The principle of bistability, non linearity or bifurication is neither exclusively of classical or quantum mechanical nature and appears in many different systems [10–13]. In a cavity QED experiment the non-linearity stems from the fact that a collection of N two-level systems is coupled to a single mode cavity. The resonator is a linear harmonic oscillator whereas the two-level spins are in a sense the most purest anhamonic system possible. Hence treating the system as coupled harmonic oscillators is no longer justified when the linear regime is left. The starting point to describe this phenomena is the Tavis-Cummings Hamiltonian as discussed earlier. From the Heisenberg operator equations $\dot{a} = \frac{i}{\hbar}[\mathcal{H}, a] - \kappa a$, $\dot{\sigma}^- = \frac{i}{\hbar}[\mathcal{H}, \sigma^-] - \gamma_\perp \sigma^-$ and $\dot{\sigma}_z = \frac{i}{\hbar}[\mathcal{H}, \sigma_z] - \gamma_\parallel \sigma_z$ and a semi-classical approach ($\langle a^\dagger \sigma_- \rangle \approx \langle a^\dagger \rangle \langle \sigma^- \rangle$) a set of first order differential equations for the operator expectation values is observed $\langle a \rangle \equiv a$, $\langle \sigma^- \rangle \equiv \sigma^-$ and $\langle \sigma_z \rangle \equiv \sigma_z$. The discussion presented below follows Ref. [5] and the equations for cavity amplitude, spin polarization and spin inversion read

$$\dot{a} = -\left[\kappa + i(\omega_c - \omega_p)\right]a + Ng\sigma_j^- - \eta \qquad (8.1)$$

$$\dot{\sigma}^- = -\left[\gamma_\perp + i(\omega_s - \omega_p)\right]\sigma_j^- - g\sigma_z a \qquad (8.2)$$

$$\dot{\sigma}_z = -\gamma_\parallel \left[1 + \sigma_z\right] + 4g\sigma^- a \qquad (8.3)$$

with cavity frequency ω_c and spin frequency ω_s. The single spin cavity coupling strength g is in enhanced by the number of spins N. Additionally a driving term with a power transmitted at a rate η and frequency ω_p is included in the cavity operator equation. In this set of equations three different dissipation rates occur: the cavity dissipation rate κ, the spin polarization decay rate given by the transverse spin relaxation time $\gamma_\perp = \frac{1}{T_2}$ and spin inversion decay rate given by the longitudinal spin relaxation time $\gamma_\parallel = \frac{1}{T_1}$. One should note that in experiments working with optical cavities often the assumption of $\gamma_\parallel = \gamma_\perp \to \gamma$ is made. However, in the presented experiments $\gamma_\parallel \gg \gamma_\perp$ which is crucial for the observed effects and can no longer be neglected. Nevertheless, the most important system parameter remain $\gamma_\perp$, κ and $\Omega = \sqrt{N}g$ which will be later used to define a dimensionless cooperativity parameter C.

The derived set of mean field Eq. 8.3 is well known as the Maxwell-Bloch equations used in the framework of non-linear optics [14]. The steady state solution and expectation values of this set of equations is found by setting the time derivations to zero. The steady state spin inversion then reads

$$\sigma_z = -\frac{1}{1 + \frac{|a|^2/n_0}{1+\Delta^2/\gamma_\perp^2}} \qquad (8.4)$$

and the steady state cavity amplitude ($|a| = \langle a \rangle$) follows as

$$|a|^2 = \frac{\eta^2}{\kappa^2} \frac{1 + \Delta^2/\gamma_\perp^2}{(1 - C\sigma_z)^2 + \Delta^2/\gamma_\perp^2} \tag{8.5}$$

with the cavity spin detuning $\Delta = \omega_s - \omega_c$ and the assumption of $\omega_c = \omega_p$. Two new characteristic dimensionless parameters are introduced: the saturation photon number

$$n_0 = \frac{\gamma_\perp \gamma_\parallel}{4g^2} \tag{8.6}$$

at which first non-linear effects occur and most importantly the cooperativity parameter

$$C = NC_0 = \frac{Ng_0^2}{2\kappa\gamma_\perp} = \frac{\Omega^2}{2\kappa\gamma_\perp}. \tag{8.7}$$

If cavity and spins are on resonance $\omega_c = \omega_s$ a very compact expression can be deduced

$$\frac{\eta^2}{\kappa^2} = |a|^2 \left(\frac{n_0 C}{|a|^2 + n_0} + 1 \right)^2 \tag{8.8}$$

linking an injected drive signal η to the created intra-cavity field amplitude $|A| \equiv \langle a \rangle$. Such a cavity spin system is then fully characterized by a cooperativity parameter C, saturation photon number n_0 and cavity dissipation rate κ.

If the number of photons in the cavity is not sufficient to saturate the spin ensemble, the suppressed cavity field intensity is given by

$$|a|^2 = \frac{\eta^2}{\kappa^2} \frac{1}{C^2} \tag{8.9}$$

which is known as the weak driving limit. If intensity of the driving field is increased the ensemble is eventually bleached and the transmitted intensity through the cavity asymptotically approaches

$$|a|^2 = \frac{\eta^2}{\kappa^2} \tag{8.10}$$

where the strong driving limit is reached. Thus operating the system between these regimes eventually gives rise to bistability and bifurcation if the cooperativity C of the system is large enough as it is shown in Fig. 8.1. How strong a bistable behavior is observable in the experiment depends naturally on the system cooperativity C. For a homogeneously broadened spin ensemble rather low values of C are sufficient to show bistability, see Fig. 8.2, but if inhomogeneouse spin broadening is present any

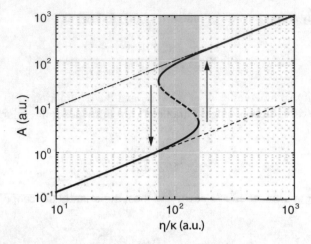

Fig. 8.1 The relation between driving field $\frac{\eta}{\kappa}$ and cavity field amplitude $|A|$ shows a bistable regime *thick black dashed line* for a extremely large cooperativity $C = 70$. In the case of low driving the cavity output is suppressed by the polarized spin ensemble and is well estimated by $|\frac{\eta}{\kappa}|^2 \frac{1}{C^2}$ *black dashed line* and in the strong driving regime *black dot dashed line* by $|\frac{\eta}{\kappa}|^2$

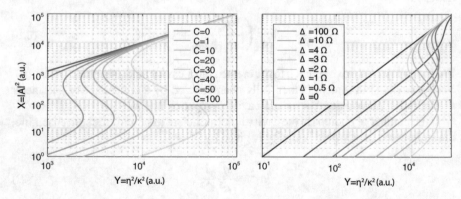

Fig. 8.2 *Left* bistability curves for different system cooperativity C for a homogeneous broadened spin ensemble on resonance with the cavity, i.e. $\omega_s = \omega_c$. *Right* As the spin ensemble is detuned from the cavity, i.e. $\Delta = \omega_c - \omega_s$ the amplitude lifting decreases. The cooperativity of $C = 100$ is kept constant while the detuning Δ is varied. In the presence of inhomogeneous spin broadening this behavior smears out any bistability

bistability might be smeared out and is disguised. Therefore the effective cooperativity has to be increased to rather large values of C. This becomes clear if Eq. 8.8 is plotted for different spin detuning Δ as is shown in Fig. 8.2.

8.3 Experiential Observation of Amplitude Bistability

As discussed earlier a possible bistable behavior is observed eventually by measuring the steady state intensity transmitted trough the cavity $|T|^2 = |\frac{A}{\eta}|^2$ as a function of the input drive power $P_{in} \propto |\eta/\kappa|^2$. As shown in Sect. 5.1.2 the dispersive cavity shift is used to estimate the longitudinal spin relaxation time $T_1 = \frac{1}{\gamma_{\parallel}} \approx 180$ s. In contrast, the transversal spin relaxation time $T_2 = \frac{1}{\gamma_{\perp}} \approx 100$ ns is approximately nine orders of magnitude smaller than T_1. These time constants and the inhomogeneous spin broadening make the experimental observation of bistability extremely difficult. The strongly coupled cavity spin system saturates only very slowly due to the extremely long T_1 time. In Fig. 8.3 the observed time constants of the cavity transmission to set into a stationary state for a given drive intensity are shown. At the point where the system sets from the low to the strong driving branch and is bleached the saturation and relaxation times are drastically increased. This means that in order to reach the stationary cavity state, great care has to be taken and a sufficiently long wait time has to be chosen to observe a stationary state for a given input power. In Fig. 8.3 also the spectral transmitted intensity through the cavity is shown for different driving powers in the stationary state. As the power level is increased the Rabi splitting vanishes and if the ensemble is bleached only the bare cavity resonance is observed as the system decouples.

In the following we will test the bistable behavior of the system by measuring the transmitted intensity through the cavity for having spin ensemble and cavity always on resonance, i.e. $\omega_s = \omega_c$, and vary the probe frequency detuning $\Delta = \omega_c - \omega_p$. In the experimental scheme the drive power $P_{in} \propto \eta^2$ is slowly increased step-wise and the corresponding and observed stationary transmission level $|T|^2$ are determined for each drive power. When the ensemble is bleached and a stationary state is reached the

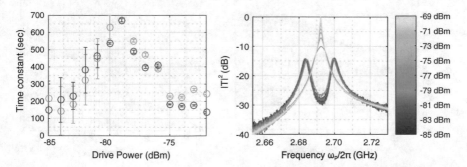

Fig. 8.3 *Left* Time constants observed by applying an exponential fit to the cavity transmission for constant drive values an a system cooperativity of $C \approx 18$. In the low and strong driving regime the time constant is determined by the longitudinal spin relaxation time $T_1 \approx 180$ s. The colors *green* and *blue* correspond to an upwards and downwards power ramp, respectively. *Right* Spectral stationary transmission for step-wise increased driving intensities. As the ensemble is bleached the Rabi splitting vanishes due to the spin saturation

drive intensity is then step-wise lowered again and the steady state transmissions are determined again. The system exhibits bistability if a regime is observable where for different drive power values different stationary transmission $|T|^2$ levels are found. In other words if two different transmission levels are found for one driving power when the intensity is ramped slowly "up" and "down" again. Which means a hysteresis or kind of memory effect should be observed in this measurement scheme.

The cooperativity of the system is determined by the collectively enhanced coupling strength $\Omega = \Omega_R/2$. This collective cooperativity can be increased by a factor of two by changing the number of spins in the cavity mode volume, as shown in Sect. 5.1.1, and allows to control the effective cooperativity. There is no direct access to $\gamma_\perp$ but by the width of the polaritonic peaks Γ and assuming $2\Gamma - \kappa = \gamma_\perp$ the effective cooperativity is estimated by

$$C = \frac{\Omega^2}{2\kappa(2\Gamma - \kappa)} . \tag{8.11}$$

One should note that due to convention the cavity linewidth is introduced as the half-width at half-maximum (HWHM), whereas Γ is denoted as the full-width at half-maximum (FWHM) linewidth. The system response is tested for two effective cooperativities by tuning two and all four NV subensembles in resonance with the cavity. For the lower collective cavity spin interaction of $\Omega/2\pi \approx \frac{\Omega_R}{4\pi} = 9.5$ MHz, a width of the polaritonic peaks $\Gamma/2\pi = 3.1$ MHz and a cavity linewidth $\kappa/2\pi = 0.44$ MHz an effective cooperativity of $C \approx 18$ is estimated. The data set corresponding to this parameter values is shown in Fig. 8.4 where no bistability is observed at all, although the cooperativity is already rather larger. In order to increase the cooperativity parameter the number of spins is changed by factor of two resulting in a

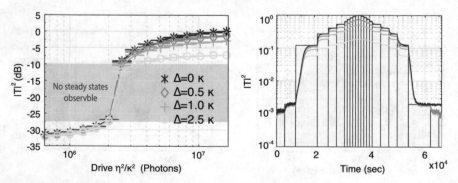

Fig. 8.4 *Left* Stationary cavity transmission for $\omega_s = \omega_c$ for different de-tunings of the cavity probe tone $\Delta = \omega_c - \omega_p$ and system parameters $\Omega/2\pi \approx \frac{\Omega_R}{4\pi} = 9.5$ MHz, $\Gamma/2\pi = 3.1$ MHz (FWHM) and $\kappa/2\pi = 0.44$ MHz (HWHM) corresponding to an effective cooperativity of cooperativity $C \approx 18$. *Right* Time traces measured to drive the system in to stationary states for different de-tunings Δ. The mean stationary transmission *left panel* is calculated from the mean transmission after a steady state was observed for each power step

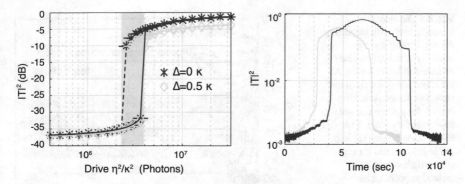

Fig. 8.5 *Left* Stationary cavity transmission for $\omega_s = \omega_c$ for different de-tunings of the cavity probe tone $\Delta = \omega_c - \omega_p$ and system parameters $\Omega/2\pi \approx \frac{\Omega_R}{4\pi} = 13.3$ MHz, $\Gamma/2\pi = 2.9$ MHz and $\kappa/2\pi = 0.44$ MHz corresponding to an effective cooperativity of $C \approx 38$. A bistable region of a ≈ 2 dB power interval is observed. *Right* Time traces measured to drive the system in to stationary states for different de-tunings Δ. The mean stationary transmission plotted in the *left panel* is calculated from the mean transmission after a steady state was observed for each power step

coupling strength of $\Omega/2\pi \approx \frac{\Omega_R}{4\pi} = 13.3$ MHz and a reduced linewidth of the polaritons $\Gamma/2\pi = 2.9$ MHz due to the cavity protection effect. With a cavity linewidth of $\kappa/2\pi = 0.44$ MHz the effective cooperativity is increased by a factor of ≈ 2.1 to $C \approx 38$. As is shown in Fig. 8.5 this parameter values are sufficient to observe amplitude bistability with a clear bistable region of two ≈ 2 dB. The drive power and stationary number of photons in the cavity at which the ensemble starts to bleach correspond to $\approx \sqrt{N} = 10^6$ with N the number of spins in the cavity mode volume. This corresponds approximately to the limit where the Holstein-Primakoff approximation of the coupled cavity spin system breaks down since the photon numbers in the cavity is larger then $\sqrt{N}$.

8.4 Conclusion

To my best knowledge this simple but very elegant measurements have been not observed in an all solid-state hybrid microwave device so far. Although demonstrated in early cavity QED experiments this measurements open up a new way in studding dynamical processes in strong non-linear medias due to the extremely long spin relaxation times. For example if the system is driven at power levels at which the system jumps from the low to the strong driving branch complex dynamical processes are observed. Such a measurement is shown in in Fig. 8.6. The system is initialized in the strong driving branch. The drive intensity is then instantaneously switched to lower drive powers. If the the stimulus power is still in the strong driving branch the steady state transmission will settle then to the corresponding drive power as is shown in Fig. 8.4. However, for drive powers at the transition between low and

Fig. 8.6 Te system is initialized far up in the strong driving branch and the drive intensity is instantaneously switched to lower levels. At the transition point from the strong to the weak driving branch the system can not set into stationary state. The cavity transmission set into the next lower lying stedy state which would be a forbidden point when looking at Fig. 8.4. However the decay towards steady state occurs on an extreme long time scale of up to 10^4 s

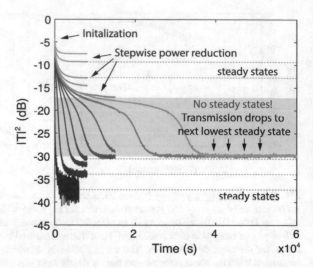

strong driving branch the system is not reaching a stationary state. As is shown in Fig. 8.6 the transmission will always drop to the next lower stationary state on extrem long time scales of up to 10^4 s. This also means that the stationary transmission reaches a value which would be forbidden as can be seen from Fig. 8.4. This an bizarre phenomenon which is not in contradiction with the semi classical Maxwell-Bloch equations, however has not been observed experimentally so far. This could find applications in understanding and testing complex bifurcation phenomenon of classical and quantum mechanical systems in the time domain.

References

1. G. Rempe, R.J. Thompson, R.J. Brecha, W.D. Lee, H.J. Kimble, Optical bistability and photon statistics in cavity quantum electrodynamics. Phys. Rev. Lett. **67**(13), 1727–1730 (September 1991)
2. C.M. Savage, H.J. Carmichael, Single atom optical bistability. IEEE J. Quantum Electron. **24**(8):1495–1498, (August 1988). ISSN 0018-9197
3. P.D. Drummond, Optical bistability in a radially varying mode. IEEE J. Quantum Electron. **17**(3):301–306, (March 1981). ISSN 0018-9197
4. P.D. Drummond, D.F. Walls, Quantum theory of optical bistability. II. Atomic fluorescence in a high-Q cavity. Phys. Rev. A. **23**(5):2563–2579, (May 1981)
5. M.J. Martin, D. Meiser, J.W. Thomsen, J. Ye, M.J. Holland, Extreme nonlinear response of ultranarrow optical transitions in cavity QED for laser stabilization. Phys. Rev. A. **84**(6):063813, (December 2011)
6. M.G. Raizen, R.J. Thompson, R.J. Brecha, H.J. Kimble, H.J. Carmichael, Normal-mode splitting and linewidth averaging for two-state atoms in an optical cavity. Phys. Rev. Lett. **63**(3), 240–243 (July 1989)
7. L.S. Bishop, E. Ginossar, S.M. Girvin, Response of the strongly driven jaynes-cummings oscillator. Phys. Rev. Lett. **105**(10), 100505 (September 2010)

8. N. Peyghambarian, H.M. Gibbs, Optical bistability for optical signal processing and computing. Opt. Eng. **24**(1):240168–240168, (1985). ISSN 0091-3286

9. P. Mandel, S.D. Smith, B.S. Wherrett, *From optical bistability towards optical computing: the European Joint Optical Bistability Project*. (North-Holland, April 1987). ISBN 978-0-444-70159-6

10. B. Shulgin, A. Neiman, V. Anishchenko, Mean switching frequency locking in stochastic bistable systems driven by a periodic force. Phys. Rev. Lett. **75**(23), 4157–4160 (December 1995)

11. R. Bonifacio, L.A. Lugiato. Cooperative effects and bistability for resonance fluorescence. Opt. Commun. **19**(2):172–176, (November 1976). ISSN 0030-4018

12. F. Brennecke, T. Donner, S. Ritter, T. Bourdel, M. Köhl, T. Esslinger. Cavity QED with a Bose–Einstein condensate. Nature. **450**(7167):268–271, (November 2007). ISSN 0028-0836

13. N. Lambert, F. Nori, C. Flindt, Bistable photon emission from a solid-state single-atom laser. Phys. Rev. Lett. **115**(21), 216803 (November 2015)

14. L. Lugiato, F. Prati, M. Brambilla. *Nonlinear Optical Systems*. (Cambridge University Press, March 2015). ISBN 978-1-107-06267-2

Chapter 9
Spin Echo Spectroscopy—Spin Refocusing

9.1 Introduction

As is shown in the previous chapter the spin ensemble can be partially excited and saturated by applying a strong microwave driving field. Therefore the spin ensemble population can be controlled coherently to some extend if the pulse intensity is strong enough. This can be used to perform spin echo spectroscopy measurements on the in homogeneously broadened spin ensemble. In such an ensemble of broadened emitters each individual spin or spin packet has distinct frequencies $\omega_{1...N}$ with a finite linewidth γ [1, 2], which we identify as the homogeneous spin life time. The sum of this transition frequencies is inhomogeneously broadened and the resulting line width is denoted as γ_{inh} as discussed in Sects. 3.3 and 5.1.2. This line broadening causes considerable spin dephasing but is a reversible process which can be refocused by a spin echo. Thus a symmetric bright spin state $|B\rangle$ dephases and propagates into a spin dark state

$$|B(t)\rangle = \frac{1}{\sqrt{N}}(|\uparrow\rangle_1 e^{-i(\omega_s-\omega_1)t} + |\uparrow\rangle_2 e^{-i(\omega_s-\omega_2)t} + \cdots + |\uparrow\rangle_N e^{-i(\omega_s-\omega_N)t}) \rightarrow |D\rangle,$$
(9.1)

hence the spin components pick up a relative phase shift during the free evolution time t. Furthermore this means that the spin component of the polariton modes $|\pm\rangle = \frac{1}{\sqrt{2}}(|1, G\rangle \pm |0, B\rangle)$ dephases and evolves into a spin dark state, hence the system decouples and the excitation is lost in the spin ensemble.

The idea of spin echos was first discussed by Erwin Hahn in 1950 [3, 4]. He showed that the decayed signal in a nuclear magnetic resonance spectrometer can be refocused by applying multiple consecutive microwave pulses. Such a pulse sequence is schematically visualized in Fig. 9.1. If a polarized spin ensemble is excited with a $\pi/2$ pulse on the equator of the Bloch sphere, the macroscopic magnetization decays with a characteristic time constant $T_2^* = 1/\gamma_{\mathrm{inh}}$, which is known as free induction decay. Therefore if the spin ensemble is brought to the equator of the Bloch sphere by a $\pi/2$ pulse, the dephased spin excitation can be refocused by a second microwave

© Springer International Publishing AG 2017
S. Putz, *Circuit Cavity QED with Macroscopic Solid-State Spin Ensembles*,
Springer Theses, DOI 10.1007/978-3-319-66447-7_9

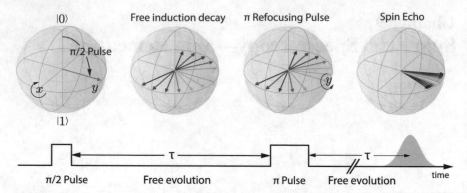

Fig. 9.1 Visualization of an idealized spin echo pulse sequence. A first $\pi/2$ pulse rotates the spin ensemble on the equator of the Bloch sphere. After the free induction decay and a wait time τ a second π pulse rotates the spins on the equator and inverts the relative acquired phase shift in the spin ensemble. After a time τ the spin ensemble refocuses and a spin echo is produced

pulse after the initial free induction decay. Therefore after the magnetization has decayed during a free evolution period τ a second π pulse fully inverts all spins on the equator, which can be understood as a time reversal. Hence a state $|B(t)\rangle$ propagated in to a dark state $|D\rangle$ evolves back into a symmetric spin bright state $(|B\rangle - \tau - |D\rangle - \tau - |B\rangle)$. This gives a spin echo after an additional wait time τ since the bright state couples to the cavity mode again. The time scale on which such a refocusing is possible is given and ultimately limited by the single spin transverse relaxation time $T_2 = 1/\gamma > T_2^*$. This simple but powerful principle has been a very important technique for many applications from magnetic resonance imaging (MRI) to quantum memory protocols [5–7].

9.2   Experimental Implementation

9.2.1   Car-Purcell-Meiboom-Gill Echo Train

A sophisticated extension of the Hahn echo technique is the so called Car-Purcell-Meiboom-Gill (CPMG) sequence as is depicted in Fig. 9.1. In such a first $\pi/2$ pulse rotates the spin ensemble along the x-axis on the equator of the Bloch sphere followed by a π pulse rotating the spins along the y-axis after a time interval τ. Instead of scanning the wait time τ many π refocusing pulses separated by an equal wait time τ are repeated which gives than the so called CPMG echo train [8]. This technique is is robust against pulse imperfections and is fast since a single sequence allows to measure the echo decay time.

In our experiments the application of a perfect π and $\pi/2$ pulse is impossible due to the inhomogeneity of the single spin-cavity interaction strengths determined by

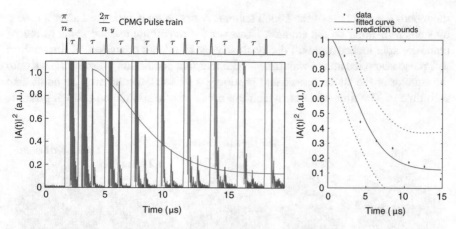

Fig. 9.2 *Left* Experimental implementation of a CPMG echo train. *Right* The CPMG echo peaks decay with a characteristic time constant $T_2^{\mathrm{CPMG}} = 4.8 \pm 1.6\ \mu$s. The confidence interval shown corresponds to the 1σ standard deviation of the fit of a stretched exponential to the observed signal

the geometry of the cavity. Therefore a pulse duration equal to the Rabi period $1/\Omega_R$ together with a high intensity pulse is used and gives the best echo fidelity. Therefore the used pulses are denoted as $\pi/n = 1/\Omega_R$ and $\pi/2n = 1/2\Omega_R$ as stimulus and refocusing pulses, respectively. We perform echo spectroscopy measurements with a magnetic field alignment of $\phi = 45°$, at which two of the four NV sub ensembles are in resonance with the cavity. This field direction results in the best achievable echo times, since the echo is sensitive to misalignment of the external d.c. magnetic field [9] with respect to the NV axis and the bath of residual electron and nuclear spins in the host material.

The experimental implementation of a CPMG sequence is shown in Fig. 9.2 in which clear echos revival are observed. The decaying and normalized echo peaks are fitted by a stretched exponential $\mathrm{e}^{-(\frac{t}{c})^b}$ with decay time c and a stretch parameter b by which the spin echo time is estimated. The best achievable echo recovering times in our experiment are $T_2^{\mathrm{CPMG}} = 4.8 \pm 1.6\ \mu$s with $b = 1.58$, which is compared to $T_2^* = \frac{1}{\gamma_{inh}} = 0.0339\ \mu$s an improvement by a factor of ≈ 130 times.

9.2.2 Stimulated Spin Echo

As originally introduced by Hahn in 1950 a spin echo is generated by two consecutive pulses of equal duration separated by a wait time τ. He also pointed out that a third echo can be produced by an additional pulse applied after an additional wait time T after the first echo, which is called a stimulated echo. In an ideal case a first $\pi/2$ pulse brings all spins to the equator of the Bloch sphere (xy plane), after the free induction decay and a wait time τ a second $\pi/2$ pulse inverts all spins and brings them onto the xz plane of the Bloch sphere. A first echo is produced after a time τ by spins remaining on the equator. However the remaining excited spins in the xz plane are spin locked and will decay slowly by T_1 related processes. If then a third $\pi/2$ pulse is applied after a wait time T, the reaming population is brought back onto the equator of the Bloch sphere and produces a stimulated echo after an additional wait time τ. The time scale on which the creation of a stimulated echo is possible

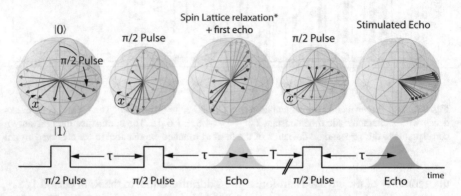

Fig. 9.3 Visualization of an idealized stimulated echo. A first $\pi/2$ pulse rotates a spin ensemble onto the equator of the Bloch sphere. After a wait time τ and the free induction decay a second $\pi/2$ moves population onto the xz plane of the Bloch sphere. After a time τ a first echo is produced. The remaining spin population decays via T_1 processes and is by a third pulse rotated back to the equator after a wait time T. This creates then a stimulated echo after a time τ

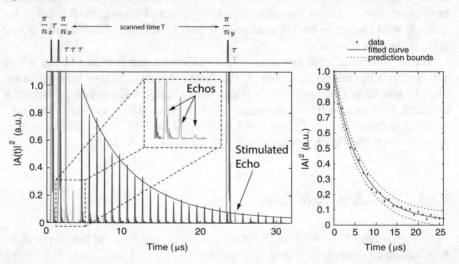

Fig. 9.4 *Left* Experimental implementation of a stimulated echo sequence. After two consecutive pulses with equal length a third one is applied after a time T. Due to the high cooperativity of the strongly coupled cavity the initial echo after the first two pulses also creates echos itself and we observe in total three echo peaks. *Right* Decay of the stimulated echo peaks with characteristic time constant $T_{1\rho} = 6.4 \pm 0.59\,\mu s$

is called the longitudinal spin lattice relaxation time in the rotating frame [1] and is commonly abbreviated as $T_{1\rho}$ (Fig. 9.3).

The experimental realization of such a sequence is shown in Fig. 9.4. The spin echo time is estimated by fitting an stretched exponential $e^{-(\frac{t}{\tau})^b}$ with decay time τ and stretch parameter b to the decaying stimulated echo peaks. The best achievable echo times in our experiment are $T_{1\rho} = 6.4 \pm 0.59 \, \mu s$ with $b = 0.96$. Due to the high cooperativity in the experiment the first echo is followed by consecutive echo peaks, this means the created echo serve as refocusing pulse for a remaining unfocused spin population.

9.3 Conclusion

This results demonstrate that in the presented experiments the echo times for stimulated and CPMG echos are approximately equal $T_{1\rho} \approx T_2^{CPMG}$. From this observation one can draw the following important conclusion on what is the limiting mechanism for the coherence time in the presented experiments: The longitudinal spin-lattice relaxation time in the rotating frame $T_{1\rho}$ is a measure of spin diffusion. Which is the characteristic time scale on which an excitation migrates or diffuses within the inhomogeneously broadened spin ensemble. This process is of course ultimately limited by T_1, but due to dipolar spin-spin interactions this time scale is drastically shortened which can be expressed by $T_{1\rho}$ and is a direct measure of the mean spin-spin coupling strength within the ensemble. In an experiment with an ideal cavity geometry such processes should be reversible by a CPMG echo and $T_{1\rho}$ should be long compared to T_2^{CPMG}. However due to the non uniform single spin Rabi frequencies, which decreases with distance from the cavity, there is a clear direction for spin diffusion in the electron spin ensemble. Therefore a collective excitation can leave the cavity mode volume by diffusion and sooner or later is lost and can not be refocused again. This manifests in our experiment with echo times $T_{1\rho} \approx T_2^{CPMG}$ and means also CPMG echos are limited by spin diffusion and not the pure spin-spin relaxation time T_2.

References

1. A. Abragam. *The Principles of Nuclear Magnetism*. (Clarendon Press, 1961). ISBN 978-0-19-852014-6
2. A. Abragam, B. Bleaney. *Electron Paramagnetic Resonance of Transition Ions*. (OUP Oxford, June 2012). ISBN 978-0-19-102300-2
3. E.L. Hahn, Spin echoes. Phys. Rev. **80**(4), 580–594 (November 1950)
4. H.Y. Carr, E.M. Purcell, Effects of diffusion on free precession in nuclear magnetic resonance experiments. Phys. Rev. **94**(3), 630–638 (May 1954)
5. W. Hua, R.E. George, J.H. Wesenberg, K. Mølmer, D.I. Schuster, R.J. Schoelkopf, K.M. Itoh, A. Ardavan, J.J.L. Morton, G. Andrew, D. Briggs, Storage of multiple coherent microwave excitations in an electron spin ensemble. Phys. Rev. Lett. **105**(14), 140503 (September 2010)

6. C. Grèzes, B. Julsgaard, Y. Kubo, M. Stern, T. Umeda, J. Isoya, H. Sumiya, H. Abe, S. Onoda, T. Ohshima, V. Jacques, J. Esteve, D. Vion, D. Esteve, K. Mølmer, P. Bertet, Multi-mode storage and retrieval of microwave fields in a spin ensemble, (2014) arXiv:1401.7939
7. W. Tittel, M. Afzelius, T. Chaneliére, R.l. Cone, S. Kröll, S.a. Moiseev, M. Sellars, Photon-echo quantum memory in solid state systems. Laser Photonics Rev. **4**(2):244–267, (February 2010). ISSN 1863-8899
8. C.P. Slichter, *Principles of Magnetic Resonance*. (Springer Science and Business Media, April 2013). ISBN 978-3-662-09441-9
9. P.L. Stanwix, L.M. Pham, J.R. Maze, D. Le Sage, T.K. Yeung, P. Cappellaro, P.R. Hemmer, A. Yacoby, M.D. Lukin, R.L. Walsworth, Coherence of nitrogen-vacancy electronic spin ensembles in diamond. Phys. Rev. B **82**(20), 201201 (November 2010)

Chapter 10
Conclusion and Outlook

The presented experimental results reveal fundamental physical effects in solid-state hybrid quantum systems. Giving an in depth treatment of macroscopic electron spin ensembles coupled to a single mode microwave cavity. The inhomogeneous spectral spin broadening has to be addressed in the framework of the Dicke and generalized Jaynes-Cummings model. These concepts gave a very intuitive insight, from which all demonstrated effects could be derived and understood. The presented discussion and experimental results therefore give the reader a detailed insight into cavity QED experiments with macroscopic ensembles of two-level systems.

In Chap. 5 it was shown how it is possible to create strongly coupled polariton modes by collective enhancement. In such large spin ensembles super and subradiant spin states have to be considered. Especially inhomogeneous broadening which lifts the degeneracy of the former dark subradiant modes, and can no longer be neglected in the discussion of the observed physical phenomena. Hence the bath of subradiant states acts as a source of decoherence and damps the polaritons modes. However, in Chap. 6 it was shown how the collective coupling strength allow the energetic decoupling of the polariton modes from the remaining bath of subradiant states by hybridization. This effect is known as "cavity protection effect" and allows to suppress spin dephasing due to spectral broadening. The fundamental limit attainable by this effect is then given by the mean of the pure dissipation rates of a single spin and the cavity interface. Therefore if the system is far enough in the strong coupling regime the inhomogeneous spin broadening can be eliminated. However, this requires rather large couplings strengths which can be difficult to be experimentally realized.

Although the ensemble is spectrally broadened the single spin coherence time can be extremely long in semiconductor crystals. However, the cavity interface adds a considerable amount of dissipation. A possible way to overcome this problem was shown in Chap. 7. Spectral hole burning allows the engineering of long-lived collective dark states. The observed coherence times are drastically improved beyond the limit attainable by the cavity protection effect. The achieved life-time of Rabi oscillation is substantially extended beyond the pure cavity decay rate and the free

© Springer International Publishing AG 2017

S. Putz, *Circuit Cavity QED with Macroscopic Solid-State Spin Ensembles*,
Springer Theses, DOI 10.1007/978-3-319-66447-7_10

induction decay time of the spin ensemble. These states are identified with collective long-lived dark states. An isolated subradiant mode is created, demonstrating that the decohrence free subspace is addressable in such experiments by spectral hole burning. For the first time this approach truly lives up to the promise of hybrid solid-state devices to perform better than its individual subcomponents.

In an ideal cavity QED experiment a resonator, a harmonic oscillator, is coupled to, the most pure anharmonic system, a two-level spin or atom. It is rather difficult to observe this anharmonicity when coupling to a macroscopic spin ensemble. These systems are normally operated in the single excitation manifold of the spin ensemble and therefore can not be distinguished from a collection of harmonic oscillators coupled to the cavity mode. Thus, a way to directly observe the spin two-level nature was presented in Chap. 8. The cooperativity of the studied system is high enough to observe amplitude bistability. This bifurcation is clear evidence of the coupled spin two-level systems and was not shown in a microwave hybrid system so far. Due to the long spin life time complex dynamical phenomena can be studied, which were not accessible in previous experiments. However, "single" spin decoherence time was not directly accessible in the presented experiments. Spin echo spectroscopic measurements are presented in Chap. 9 to estimate the relevant spin coherence times of the system. The performed CPMG and stimulated spin echo measurements allowed the conclusion that the system coherence time is limited by spin diffusion. This is due to the specific geometry of the cavity. The single spin Rabi frequencies are very inhomogeneous, which means an excitation can diffuse outwards and leave the cavity mode volume.

The presented experiments clearly give insight into the complex dynamics of solid-state hybrid quantum devices. The presented approach opens up the way for long-lived quantum memories, solid-state microwave frequency combs [1], optical to microwave quantum transducers [2, 3], superradiance [4], spin squeezed states [5] and the generation of non-classical spin states in solid-state macroscopic spin ensembles [6, 7]. In order to implement these experiments a major problem in the presented setup has to be addressed and solved. The main bottleneck in the devices studied are the inhomogeneous single spin Rabi frequencies. Due to the cavity geometry, of the transmission line resonator, the single spin interaction strength is spatially dependent. This inhomogeneity makes it impossible to uniformly and coherently manipulate the ensemble spin population. A symmetric spin state will dephase if the single excitation manifold is left. Therefore instead of climbing the Dicke ladder on symmetric spin states, the system will be driven into the subradiant dark subspace.

A possible way to overcome this issue is to use three dimensional [8–10] microwave cavities. As a matter of fact the geometry of these cavities allows the creation of very homogeneous field distributions inside the resonator. However the cavity mode volumes are rather large which cause very low single spin interaction strengths. Therefore the ensemble size would need to be extremely large. A trick to resolve this problem is to focus the magnetic field inside the cavity. This is possible by adding a structure inside the resonator, creating additional capacitors and altering the magnetic and electric field distribution. The achievable single spin coupling strengths are comparable with the coupling strength attainable by transmission line

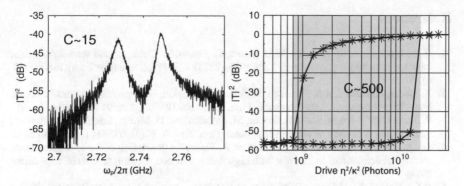

Fig. 10.1 Bistability measurement in an ultra dense spin ensemble with a concentration of ≈ 60 ppm NV. *Left* From the Rabi splitting a cooperativity of $C \approx 15$ is estimated. *Right* Amplitude bistability measurement showing an extremely large bistable area of 10 dB. This is in contradiction to the Rabi splitting and reveals the first phenomena caused by strong spin spin interactions

resonators. Additionally the cavity mode volume is decreased since the structure changes the cavity dispersion relation. This approach shows first promising results and I am convinced that with this device the performance will be greatly improved. The first experiments could allow one to realize true superradiance in a macroscopic spin ensemble, the implementation of the Dicke lattice model [11] and diamond masers [12, 13]. Additionally this approach would allow true measurements of NV spin relaxation times T_1 at millikelvin temperatures, which is an outstanding task and could help to get a better insight into ensemble dynamics at cryogenic temperatures. Moreover this approach allows open cavity structures which makes them ideal candidates to interface cold atomic clouds and combine them with superconducting qubits [14].

This approach also shows a route to a new class of cavity QED experiments with ultra dense spin ensembles, where dipole spin-spin interactions become important and many-body phenomena will be directly accessible on a chip. In a first experiment a diamond crystal containing ≈ 60 ppm NV centers was used. At such high concentrations the dipole spin-spin interaction is ≈ 1 MHz and in the range of the cavity linewidth. Therefore these strong interaction can no longer be neglected. First signatures of these strong spin-spin interactions have been observed in experiments measuring the amplitude bistability. A huge bistable regime of 10 dB was observed as is shown in Fig. 10.1. From this hysteresis a huge cooeprativity of $C \approx 500$ would have to be expected. However, the system cooperativity observed by resolving the Rabi splitting shows only a rather low cooperativity of $C \approx 15$. This deviation is counter intuitive and is identified with the large spin-spin interaction strength within the ensemble. To determine the exact mechanism causing this result has to be explored in future experiments. Nevertheless, this experiments pave the way for the implementation of experiments beyond the standard Dicke and Tavis-Cummings model.

References

1. H. de Riedmatten, M. Afzelius, M.U. Staudt, C. Simon, N. Gisin, A solid-state light–matter interface at the single-photon level. Nature, **456**(7223):773–777, (December 2008). ISSN 0028-0836
2. K. Stannigel, P. Rabl, A.S. Sørensen, P. Zoller, M.D. Lukin, Optomechanical transducers for long-distance quantum communication. Phys. Rev. Lett. **105**(22), 220501 (November 2010)
3. S. Blum, C. O'Brien, N. Lauk, P. Bushev, M. Fleischhauer, G. Morigi, Interfacing microwave qubits and optical photons via spin ensembles. Phys. Rev. A. **91**(3), 033834 (March 2015)
4. J.A. Mlynek, A.A. Abdumalikov, C. Eichler, A. Wallraff, Observation of Dicke superradiance for two artificial atoms in a cavity with high decay rate. Nat, Commun. **5**, 5186 (November 2014)
5. D.J. Wineland, J.J. Bollinger, W.M. Itano, D.J. Heinzen, Squeezed atomic states and projection noise in spectroscopy. Phys. Rev. A. **50**(1), 67–88 (July 1994)
6. M.H. Schleier-Smith, I.D. Leroux, Vladan Vuletić. Erratum: squeezing the collective spin of a dilute atomic ensemble by cavity feedback. [Phys. Rev. A. textbf81, 021804(R) (2010)]. Phys. Rev. A. **83**(3):039907, (March 2011)
7. R. McConnell, H. Zhang, J. Hu, S. Ćuk, V. Vuletić. Entanglement with negative Wigner function of almost 3,000 atoms heralded by one photon. Nature, **519**(7544):439–442, (March 2015). ISSN 0028-0836
8. H. Paik, D.I. Schuster, L.S. Bishop, G. Kirchmair, G. Catelani, A.P. Sears, B.R. Johnson, M.J. Reagor, L. Frunzio, L.I. Glazman, S.M. Girvin, M.H. Devoret, R.J. Schoelkopf, Observation of high coherence in josephson junction qubits measured in a three-dimensional circuit QED architecture. Phys. Rev. Lett. **107**(24), 240501 (December 2011)
9. S. Probst, A. Tkalčec, H. Rotzinger, D. Rieger, J.M. Le Floch, M. Goryachev, M.E. Tobar, A.V. Ustinov, P.A. Bushev, Three-dimensional cavity quantum electrodynamics with a rare-earth spin ensemble. Phys. Rev. B. **90**(10), 100404 (2014)
10. D.L. Creedon, J.-M. Le Floch, M. Goryachev, W.G. Farr, S. Castelletto, M.E. Tobar, Strong coupling between $P1$ diamond impurity centers and a three-dimensional lumped photonic microwave cavity. Phys. Rev. B. **91**(14), 140408 (April 2015)
11. L.J. Zou, D. Marcos, S. Diehl, S. Putz, J. Schmiedmayer, J. Majer, P. Rabl, Implementation of the dicke lattice model in hybrid quantum system arrays. Phys. Rev. Lett. **113**(2), 023603 (July 2014)
12. K. Sandner, H. Ritsch, R. Amsüss, Ch. Koller, T. Nöbauer, S. Putz, J. Schmiedmayer, J. Majer, Strong magnetic coupling of an inhomogeneous nitrogen-vacancy ensemble to a cavity. Phys. Rev. A. **85**(5), 053806 (May 2012)
13. L. Jin, M. Pfender, N. Aslam, P. Neumann, S. Yang, J. Wrachtrup, R.-B. Liu, Proposal for a room-temperature diamond maser. Nat. Commun. **6**, 8251 (September 2015)
14. J. Verdú, H. Zoubi, Ch. Koller, J. Majer, H. Ritsch, J. Schmiedmayer, Strong Magnetic Coupling of an ultracold gas to a superconducting waveguide cavity. Phys. Rev. Lett. **103**(4), 043603 (July 2009)

Curriculum Vitae

Stefan Putz

Born: Gmunden, 1983
Citizenship: Austria
Email: stef.putz@gmail.com

Present Address:
Department of Physics, Jadwin Hall
Princeton University

Education

- Dr. tech. (PhD) with distinction, TU Wien March 2016
  Thesis: Cavity QED with macroscopic spin ensembles
  Advisor: Johannes Majer and Jörg Schmiedmayer
- Diplom Ingenieur (M.S.) with distinction, TU Wien April 2011
  Thesis: Coherent Manipulation of Nitrogen-Vacancy Center Electron Spins
  Advisor: Jörg Schmiedmayer
- Matura (high school) with honours, Höhere technische Lehranstalt HTL1 Linz,
  2001

© Springer International Publishing AG 2017
S. Putz, *Circuit Cavity QED with Macroscopic Solid-State Spin Ensembles*,
Springer Theses, DOI 10.1007/978-3-319-66447-7

Appointments

- Postdoctoral Research Associate, Princeton University, April 2016
- Doctoral Program Building Solids for Function FWF-Solids4Fun TU Wien, 2012

Prices & Honors

- Springer Thesis Award recognizing outstanding Ph.D. research, May 2017

Academic Services

- Reviewer for the Annals of Physics, EPJ Quantum Technology, npj Quantum Information

Research stays/collaborations

- NTT Basic Research Labs, Tokyo Japan
- NII National Institute of Informatics, Tokyo Japan
- Humboldt University Berlin (Erasmus exchange)

Printed in the United States
By Bookmasters